INTRODUCTION TO MICROBIOLOGY

A.S. RAO
Formerly Professor of Botany
Nagarjuna University

PHI Learning Private Limited
Delhi-110092
2025

*In fond memory of **Shri Asoke K. Ghosh** (October 1942 – February 2024), Founder Chairman and Managing Director of PHI Learning, whose vision endlessly inspires.*

The Legacy Continues....

Published by Pushpita Ghosh, PHI Learning Private Limited, Rimjhim House, 111, Patparganj Industrial Estate, Delhi-110092 and Printed by Syndicate Binders, A-20, Hosiery Complex, Noida, Phase-II Extension, Noida-201305 (N.C.R. Delhi).

₹525.00

INTRODUCTION TO MICROBIOLOGY
A.S. Rao

ISBN-978-81-203-1217-3 (Print Book)
ISBN-978-93-5443-923-0 (e-Book)

CONTENTS

Preface v

Introduction **1–4**

1. The Structure of Microbial Cells **5–16**

Prokaryotic Cell 7
Cytoplasm of Eukaryotes 11
Cell Division 15

2. A Survey of Microorganisms and Viruses **17–28**

Prions 17
Viroids 17
Viruses 18
Rickettsias 18
Mycoplasmas 18
Bacteria 20
Cyanobacteria 21
Prochlorophyta 21
Protozoa 21
Algae 23
Fungi and Slime Moulds 23

3. Morphology and Classification of Bacteria **29–41**

Morphological Features 29
Colony Characters 30
Physiological and Genetic Characters 31
Numerical Taxonomy 31
Major Groups of Bacteria 31

4. Viruses **42–60**

Symmetry of Viruses 42
Transmission of Viruses 44
Composition of Viruses 44
Serology of Viruses 48
Replication of Viruses 49
Laboratory Study of Viruses 49
Plant Viruses 51
Animal (and Human) Viruses 52

5. **Nutrition and Growth of Bacteria** **61–67**

Sources of Nutrients 61
Entry of Nutrients into the Cell 63
Factors Affecting Microbial Growth 64
Growth of Bacteria 66

6. **Metabolism of Bacteria** **68–107**

Enzymes 68
Oxidation-reduction 69
Electron Carriers 71
Photosynthetic Pigments and Leghaemoglobin 72
Metabolism in Bacteria 73
Respiration 73
Bacterial Fermentations 84
Biosynthesis of Cellular Compounds 89
Metabolism of Autotrophs 100
Biological Nitrogen Fixation 103

7. **Microbial Genetics** **108–123**

Mutations 108
Genes and Genetic Code 111
Recombinations in Bacteria and Phages 117
Recombinant DNA 121

8. **Immunology** **124–135**

Cells of the Immune System 124
Antigen Recognition and Antibody Formation 126
Acquired Immunity 132
Hypersensitivity and Allergies 133
Serology 133

9. **Microorganisms in the Environment** **136–144**

Soil Microorganisms 136
Microorganisms in Aquatic Habitats 142
Microorganisms and Pollution 142
Microorganisms in Sewage 143
Methane Production 144

10. **Microorganisms in Industry and Food** **145–164**

Raw Materials for Fermentation 146
Fermentation Processes 148
Products of Industrial Fermentations 152
Milk Products 158
Microbiology of Perishable Foods 159

11. Microorganisms in Human Diseases **165–180**

Normal Microflora of the Human Body 166
General Account of Pathogenic Bacteria 167
Diseases Caused by Bacteria 168
Diseases Caused by Fungi 174
Diseases Caused by Protozoa 175
Epidemiology and Control of Diseases 177
Chemotherapy 179

12. Basic Methods in Microbiology **181–189**

Microscopic Methods 181
Techniques of Sterilization 184
Media Preparation 185
Isolation and Inoculation 186
Direct Observation and Staining Techniques 186
Maintenance and Preservation of Cultures 188

List of Books for Further Reading *191–192*

Index *193–207*

PREFACE

The scope of microbiology is expanding rapidly, providing information in fundamental and applied aspects in frontier areas of genetic engineering, environmental science, cancer studies and industrial processes, besides the traditional areas of medicine and agriculture. Extensive research using microorganisms is being conducted at numerous laboratories around the world, by scientists and technologists trained in various disciplines. It has thus become essential to impart and propagate knowledge in this area of research and application in a systematic manner improving over the current methods of instruction.

Microbiological studies have still not reached an advanced stage in India. However, the increased importance given to microbiology in universities and colleges has resulted in the starting of several new departments with the singular purpose of imparting undergraduate and postgraduate instructions. This expansion is being implemented rapidly as full-fledged courses in our colleges and universities, necessitating the provision of suitable texts and research sources.

There are very few books published in India which are used as sources of recent information for students and teachers of microbiology. In particular, teachers rely on books published abroad to cover various aspects of the subject. These books are limited in availability and too expensive to the average student.

This book is aimed at students and teachers and covers various aspects of the subject, emphasizing fundamental concepts with suitable examples. A wide range of topics have been dealt with and a large number of diagrams are provided to illustrate the text. The book has been kept concise in order to make it affordable. The authenticity of the covered material has been carefully verified. It is hoped that this book will be useful as a text for both undergraduate and postgraduate students of microbiology.

I wish to express my sincere gratitude to my colleagues, Prof. K.V. Malliah and Dr. M. Vijayalakshmi and my research associates, Dr. K. Ammani, Dr. V. Umamaheswar Rao and Dr. M. Raghuram, for their help in providing source material, for useful discussions and for going through the manuscript. My thanks are due to Prof. D. Santhakumari, Head of the Department of Botany, Nagarjuna University for encouragement provided by her. I am grateful to the University Grants Commission for awarding me Emeritus Fellowship which enabled me to prepare the manuscript. I wish to thank Mr. G. Sai Babu of the Vignan Computer

Centre for providing Computer resources. Finally, I thank my wife, Suseela, for her understanding and support throughout the development of this book.

Any suggestions for improving the contents will be warmly appreciated.

A.S. RAO

INTRODUCTION

Microorganisms are widespread in nature and participate in a number of processes that ensure the continuation of life on earth. They contribute to soil fertility, cleansing of the environment, and are employed in many industries connected with foods and medicines. Some microorganisms, however, cause diseases in other organisms including man, animals and plants. They have become useful tools in the study of molecular biology and genetics, and are hence, being studied by scientists and technologists trained in various disciplines.

Recognition of the existence of microorganisms dates back to 1673, when Anton van Leeuwenhoek (1632–1723), a Dutch businessman, who made lenses as a hobby, discovered the presence of minute organisms through a simple magnifying device he had made. He called the organisms *animalcules*. From his descriptions it appears that what he saw from deposits on teeth were bacteria and protozoa. However, several years passed before their significance was recognized. There had been a controversy about whether various forms of life arose spontaneously or only from preexisting life. Lazzaro Spallanzani (1729–1799), an Italian naturalist showed that meat which was heated in flasks and kept sealed did not get spoiled. This was contested by the advocates of the spontaneous generation theory. Finally the controversy was resolved when Louis Pasteur (1822–1895) devised suitable experiments to show that it was the dust in air which carried germs that caused spoilage. He followed these with demonstrations highlighting the role of microorganisms in causing fermentation. It was already known that microorganisms are associated with certain diseases. In 1835, Agostino Bassi showed that a fungus was associated with a silkworm disease. Berkeley showed in 1845 that potato blight, which caused the great potato famine in Ireland was caused by a fungus. Joseph Lister (1827–1912) recognized the role of microorganisms in human diseases and developed antiseptic techniques for surgical operations.

The first direct evidence for the role of bacteria in disease was provided by the work of Robert Koch (1843–1910) who showed that *Anthrax* disease of sheep and humans was caused by the bacterium *Bacillus anthracis.* He had investigated the disease thoroughly, isolated the pathogen and confirmed the bacterium as the causative agent. He proposed the following criteria, now known as **Koch's postulates,** to prove the causal relationship between the microorganism and the host (Koch, 1884).

- The microorganism must be present in every case of disease but absent from healthy individuals.

- The suspected microorganism must be isolated and grown in pure culture.
- The same disease must result when the organism in culture is introduced into a healthy individual.
- The same microorganism must be re-isolated from the diseased individual.

Koch developed semisolid nutrient media (with gelatin, and later with agar agar) for culturing microorganisms. In 1882, Koch isolated the organism that caused tuberculosis. The work of Pasteur and Koch greatly stimulated the study of microorganisms. The development of powerful microscopes had aided the study greatly. To separate microorganisms from body fluids, porcelain candles (known as **Chamberland filters,** after the discoverer) were developed during that period. This has facilitated the isolation and study of microorganisms in laboratory cultures.

The investigations of Pasteur between 1857 and 1860, on the role of yeasts and bacteria in various fermentations, led to the development of **pasteurization** for preventing spoilage of wine. Pasteur also showed that fermentations were carried out under anaerobic conditions. This led to the recognition that life exists in the absence of oxygen also.

The role of microorganisms in natural environments owes its recognition to the works of Martinus Beijerinck (1851–1931) and Sergei N. Winogradsky (1885–1953). Beijerinck made fundamental contributions to the ecology of microorganisms. He had isolated *Azotobacter*, a nitrogen-fixing organism, from soil and also the root nodule bacterium *Rhizobium* from leguminous plants. Winogradsky had isolated bacteria from soil which oxidized ammonia, sulphur, and iron. Their importance in biogeochemical cycles of nutrient elements, particularly carbon, nitrogen and sulphur in nature has been well recognized now.

The vaccination against smallpox was developed by Jenner (1798) but the nature of its action remained unknown for a long time. Pasteur had developed a **vaccine** against rabies by passing the fluid from a rabid dog into a rabbit repeatedly and using the rabbit serum as vaccine. Towards the close of the nineteenth century, some disease-causing bacteria were shown to produce toxins. Behring and Kitosano independently showed that inactivated toxin of the diphtheria bacillus, when injected into rabbits, yielded serum from the latter that acted as a vaccine. This study provided the basis for the understanding of the immunity principle and resulted in subsequent development of **immunology,** an important branch of medicine dealing with disease control and organ transplants.

The existence of virus diseases was established when Ivanowsky (1891), a Russian scientist found that juice from a tobacco plant with mosaic symptoms, in spite of passing through filter candles that hold back bacteria still caused the disease. Beijerinck then correctly interpreted that the disease was caused by an agent much smaller than bacteria and called it a *virus*. Their pioneering studies have led to the recognition of viruses as disease-causing agents. By the end of the nineteenth century, the causal agents of several infectious diseases had been discovered.

The early part of the twentieth century saw the extension of studies which showed the ubiquitous occurrence of microorganisms in diverse habitats in nature, their role in soil fertility and their importance as agents of disease in man, animals and plants. Two discoveries during this period had profound influence on the development of microbiology. One was the discovery of *bacteriophages* and the other the discovery of *penicillin.* The bacteriophages were observed independently by Twort and deHerëlle during 1915-17. At that time it was hoped that those viruses which affect bacteria could be used to control bacterial diseases, but this proved elusive. They were, on the other hand, subsequently shown to be very useful in genetic studies that led to our understanding of the life processes. The discovery of several antibiotics, following the discovery of penicillin has revolutionized the practice of **chemotherapy,** i.e., curing of diseases by chemicals, and contributed greatly to the health of people.

The development of sophisticated techniques such as electron microscopy, ultracentrifugation and radioactive tracer methods made it possible to probe the physical and chemical nature of bacteria and viruses. In 1936, Wendel Stanley crystallized the tobacco mosaic virus and showed it to be a particle. Bawden and Pirie (1937) showed that it is made up of a nucleic acid and a protein. Earlier, Griffith (1928) showed that a bacterium causing pneumonia could be transformed by exposure to a related strain. In 1944, Avery et al. showed that the transforming principle is a nucleic acid. The above discoveries have laid the foundation for the development of **molecular genetics.** Viruses and bacteria, because of their simple structure and rapid multiplication, have proved to be invaluable tools in biochemical and genetic studies.

Advances in genetic engineering utilizing microorganisms and viruses has resulted in the development of highly efficient industrial microorganisms leading to cheaper production of life-saving drugs such as antibiotics and insulin. The growth of a new branch of applied science i.e. **biotechnology** owes much to the study of microorganisms. Table 1 lists important events in the development of microbiology.

The importance of microorganisms in cleansing the environment is also attracting wider attention as their role in detoxifying harmful chemicals and releasing nutrient elements is better understood now. The spread of print and electronic media is contributing to public awareness of the importance of microorganisms in nature and on man's well-being. This is reflected in increasing popularity of microbiology as a subject of study at college and university levels. The present account aims at briefly presenting the elements of the structure and functions of microorganisms and it is hoped that it would be useful to students and teachers alike at the college and university levels.

Table 1 Important Events in the Development of Microbiology

1676	Leeuwenhoek discovers animalcules
1798	Jenner develops vaccine against smallpox
1799	Spallanzani disputes Spontaneous generation theory
1838	Schleiden and Schwann propose Cell theory
1857	Pasteur discovers lactic acid bacteria
1861	Pasteur disproves spontaneous generation theory
1876	Koch establishes that anthrax is caused by a bacterium
1884	Kochs publishes his postulates
1884	Gram develops staining of bacteria
1885	Pasteur develops rabies vaccine
1887	Winogradsky discovers autotrophic bacteria
1888	Beijerinck isolates bacteria from root nodules
1892	Ivanowsky studies the cause of tobacco mosaic
1899	Beijerinck establishes that tobacco mosaic is caused by virus
1915	de Herelle and Twort discover bacteriophages
1928	Griffith discovers transformation in bacteria
1929	Fleming discovers penicillin
1935	Stanley crystalizes tobacco mosaic virus particle
1941	Beadle and Tatum propose 1 gene-1 enzyme hypothesis
1944	Avery, et al. show that transformation is due to DNA
1944	Waksman discovers streptomycin
1946	Lederberg and Tatum describe bacterial conjugation
1952	Hershey and Chase show that DNA of phage enters the host cell
1952	Zinder and Lederberg discover transduction
1955	Jacob and Monod discover F-factor plasmid in *E. coli*
1961	Jacob and Monod propose operon concept
1964	Nierenberg and Khorana elucidate gene concept
1970	Arber and Smith discover endonucleases
1970	Temin and Baltimore discover reverse transcriptase
1975	Kohler and Milstein develop monoclonal antibodies
1983	Gallo and Montagnier discover HIV
1984	Mullis discovers polymerase chain reaction
1987	Feldman develops transgenic plants with *Agrobacterium* T-DNA
1992	Oliver et al. sequence yeast chromosome III completely

1

THE STRUCTURE OF MICROBIAL CELLS

The diversity of living organisms makes it imperative to classify them into groups, so that their identity can be established. **Taxonomy** is the branch of biology dealing with classification. The study of characterizing and identifying organisms is also known as **systematics**. The list of rules governing the method of classification is known as **code of nomenclature**. Societies of biologists ·dealing with major groups of organisms meet periodically and determine the rules. Such codes are referred to as **International codes of nomenclature**. Application of the nomenclatural rules is helpful not only to place any given organism in the classification system, but also to describe and name any organism that is newly discovered.

To facilitate classification, the taxonomic units are divided into ranks in a hierarchial system, each larger group being made up of a number of smaller groups. All individuals which are fairly uniform in character and compatible in reproduction are regarded as one **species.** The species is the basic unit of taxonomy. Next in hierarchy is **genus** which is a collection of closely resembling species. Similarly, genera are grouped into **families**, families into **orders**, orders into **classes**, classes into **phylae,** and phylae into **kingdoms.** Each organism is designated by the name of the genus followed by that of the species. This is known as the **binomial system of nomenclature** and it is useful for uniformity in fixing the identity of the organisms.

In all unicellular or multicellular organisms, a cell is the basic functional unit capable of carrying out all life processes. The majority of biologists now accept the division of cellular organisms into the plant, animal, and protistan kingdoms. Microorganisms constitute the protistan kingdom and are commonly referred to as **protists.** There are some living organisms which do not have a cellular structure but exhibit several properties of life. Viruses and related forms fall under this category. They are also studied along with microorganisms, as they share some common properties, especially as disease-causing agents, and many methods of their study are common. Before taking up a consideration of the taxonomy of protists, it would be advantageous to learn about basic cell structure.

All cellular organisms are divided into *prokaryotes* and *eukaryotes* on the basis of the type of cells they possess. The cell of prokaryotes is simpler and usually small (average diameter, one micron) and does not possess a distinct nucleus, whereas the eukaryotic cell is larger (5 to 10 microns or more) and more complex, possessing a distinct nucleus and subcellular organelles such as **plastids** and **mitochondria**. Figure 1 illustrates the basic features of prokaryotic and eukaryotic cells.

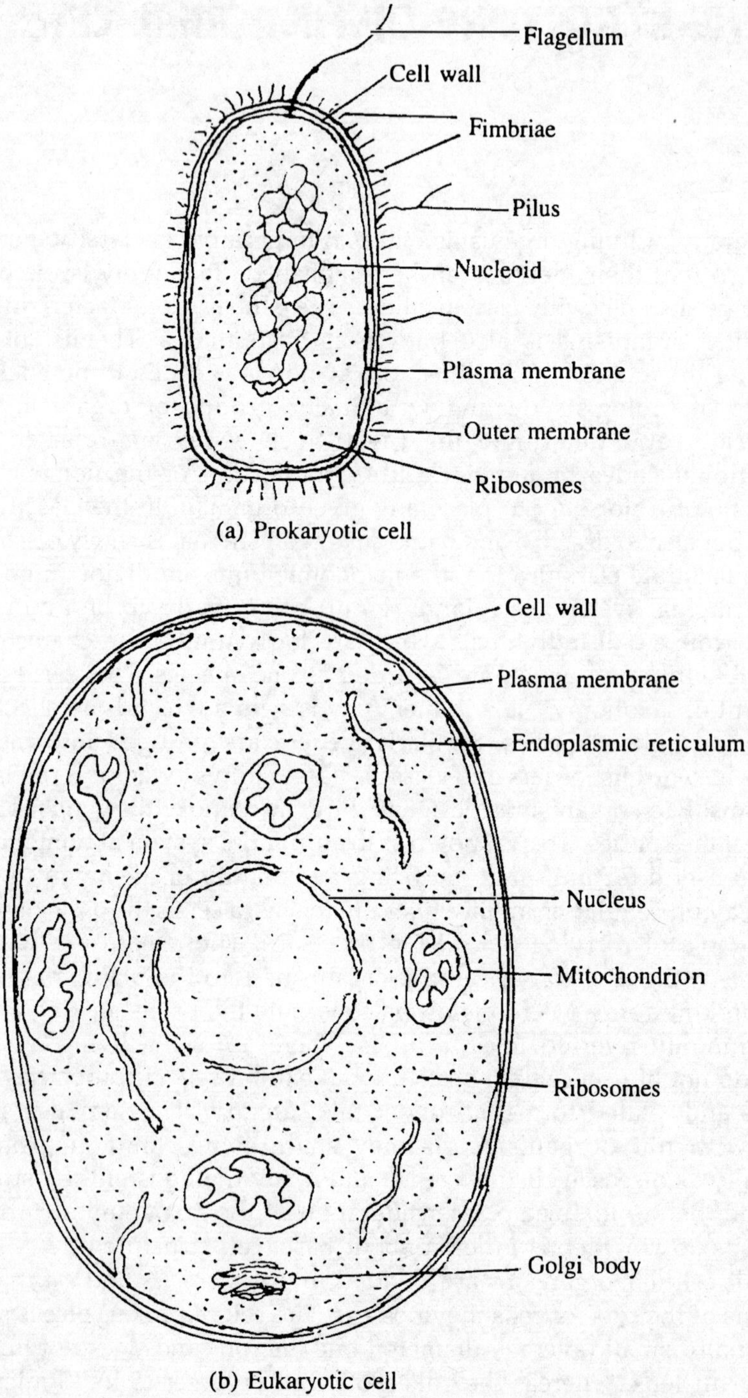

Fig. 1 General features of prokaryotic and eukaryotic cells.

PROKARYOTIC CELL

The protista are divided into eubacteria, archaebacteria and eukaryotes. The first two groups are largely unicellular and the cells are enclosed in rigid cell walls. Previously both were grouped under the phylum 'bacteria' but are now separated on the basis of recent findings on cell wall chemistry and composition of plasma membrane of the two groups. The distinguishing features of the cells of the three groups are listed in Table 2.

Table 2 Distinguishing Features of Cells in Major Groups of Microorganisms

	Eukaryotes	*Eubacteria*	*Archaebacteria*
Cell structure	Eukaryotic	Prokaryotic	Prokaryotic
Typical diameter	10 μm	1 μm	1 μm
Cell membrane			
(a) Major lipids	Glycerol diesters	Glycerol diesters	Glycerol diethers or diglycerol tetraethers
(b) Side chain	Fatty acids	Fatty acids	Polyisoprenoid alcohols
(c) Other	Steroids common	Steroids rare	Steroids rare
Internal membrane systems			
(a) Nuclear membrane	Present	Absent	Absent
(b) Endoplasmic reticulum	Present	Absent	Absent
(c) Mitochondria	Present	Absent	Absent
(d) Golgi apparatus	Present	Absent	Absent
(e) Chloroplasts	Present in phototrophs	Absent	Absent
Endo/exocytosis	Present	Absent	Absent
Cell wall contents	Peptidoglycan absent	Peptidoglycan present	Peptidoglycan absent
Flagella	9 + 2 micro-tubules	1 – 2 filaments	Microtubules
Genetic systems			
(a) Number of chromosomes	More than one	One	One
(b) Arrangement of DNA	Linear	Circular	NA
(c) Chromosome segregation	Mitotic spindle	Cell membrane	NA
(d) Introns	Usually present	Absent	Occasionally present
(e) Meiosis	Present	Absent	Absent
(f) Ribosome size	80 S	70 S	70 S
Sexual Reproduction	Frequent	Rare and often incomplete	NA

The Gram staining reaction is one of the first tests applied in studying bacteria. The reaction is either positive (Gram +) or negative (Gram –). This difference is due to the differences in the nature of cell wall of the two groups. In Gram-positive eubacteria, the wall is made up of a thick layer of **peptidoglycan**, (40/50 nm) enclosing the plasma membrane. In Gram-negative eubacteria, the peptidoglycan layer is thin (1–3 nm) and there is an outer membrane enclosing the cell. This outer membrane is made up of **lipopolysaccharides** and **lipoproteins**. The area between the outer membrane and the plasma membrane is known as **periplasmic space**. Figure 2 illustrates the difference in wall structures of Gram+ and Gram– eubacteria.

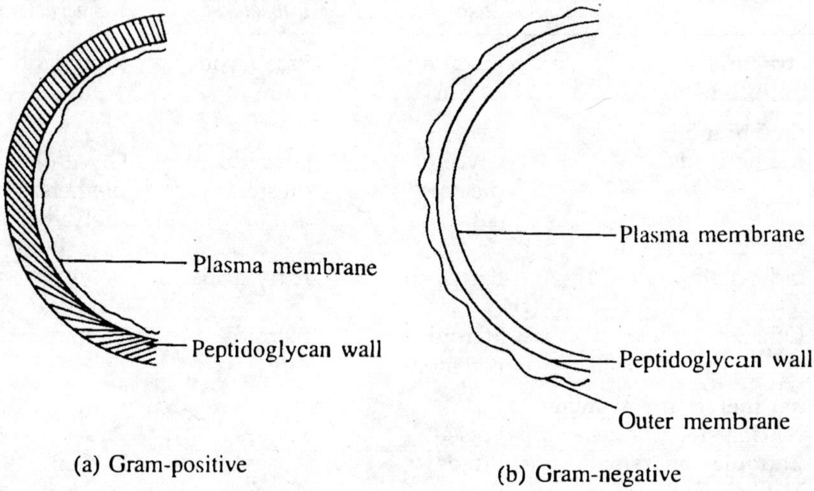

Fig. 2 Diagram showing differences in cell walls of Gram-positive and Gram-negative bacteria.

The peptidoglycan layer (also known as **murein**) is made up of a framework of polymers of two sugar derivatives, N-acetylmuramic acid and N-acetylglucosamine, and several amino acids including D-glutamic acid, L-alanine and meso-diaminopimelic acid (amino acids not found in proteins). The sugars are linked by β-1, 4 links and connected by peptide cross links. In Gram-positive bacteria, the thick peptidoglycan layer shows peptide interbridges. Polymers known as **teichoic acids** are also present in them. The arrangement of various wall components of both Gram+ and Gram– bacteria is illustrated in Fig. 3. In archaebacteria, muramic acid and D-amino acids (characteristic of eubacteria) are absent. Those giving Gram-negative reaction have a **glycoprotein** or **protein** making up the wall and those giving Gram-positive reaction have a thicker and homogeneous wall.

Plasma Membrane

The plasma membrane of eubacteria is a typical **unit membrane**, 4-5 nm thick lipid bilayer described as a **fluid mosaic model.** The lipids are made up of straight-chain molecules with hydrophilic ends towards outer and

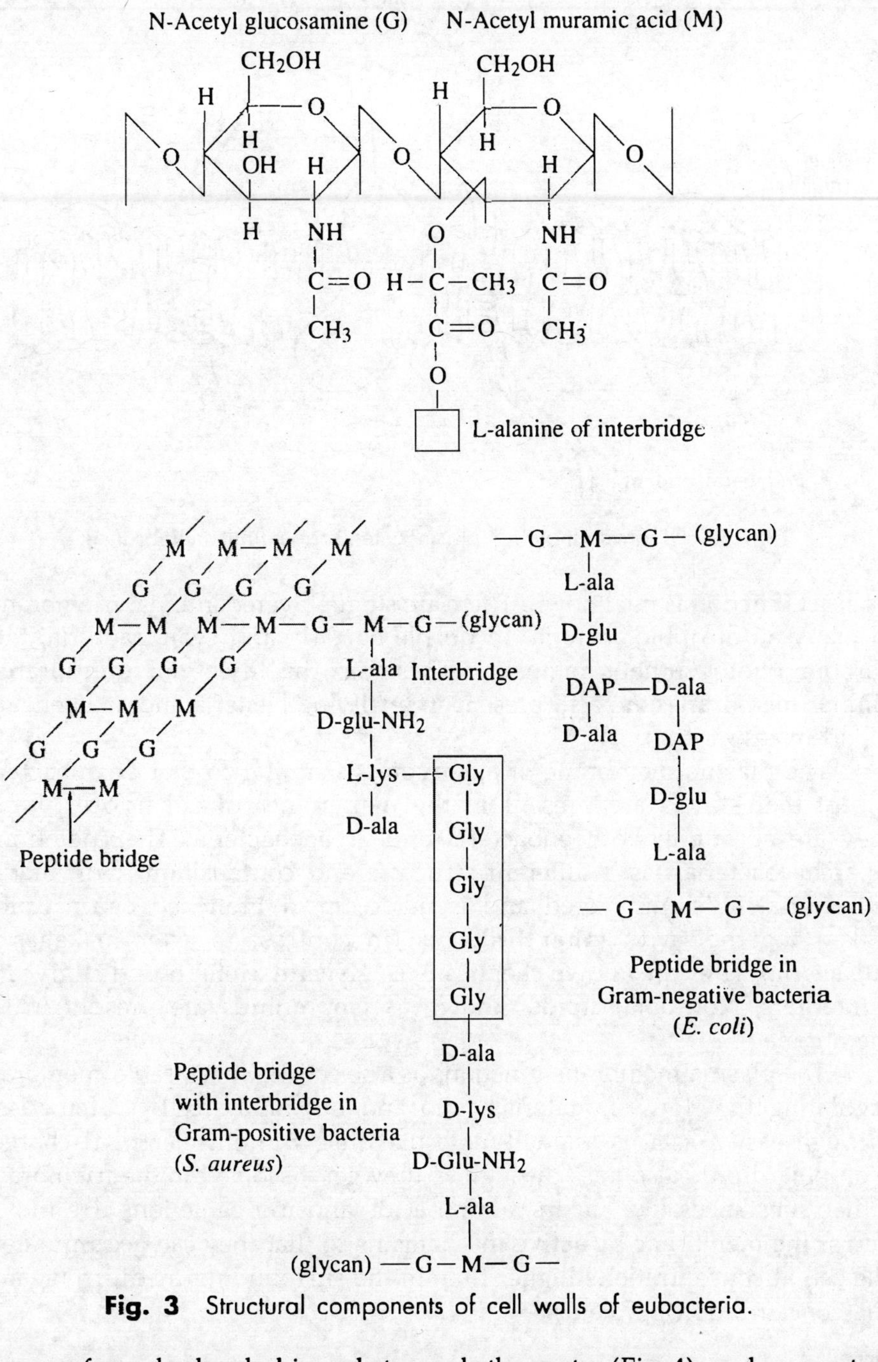

Fig. 3 Structural components of cell walls of eubacteria.

inner surfaces, hydrophobic ends towards the centre (Fig. 4), and connected by **ester linkages**. There are peripheral proteins (loosely connected to the lipids) and **integral proteins** embedded in the membrane. Some of them extend across the lipid bilayer while others go only into a part of the layer.

In many Gram-positive bacteria, and some Gram-negative ones, the plasma membrane shows inward folding, which is known as **Mesosome**.

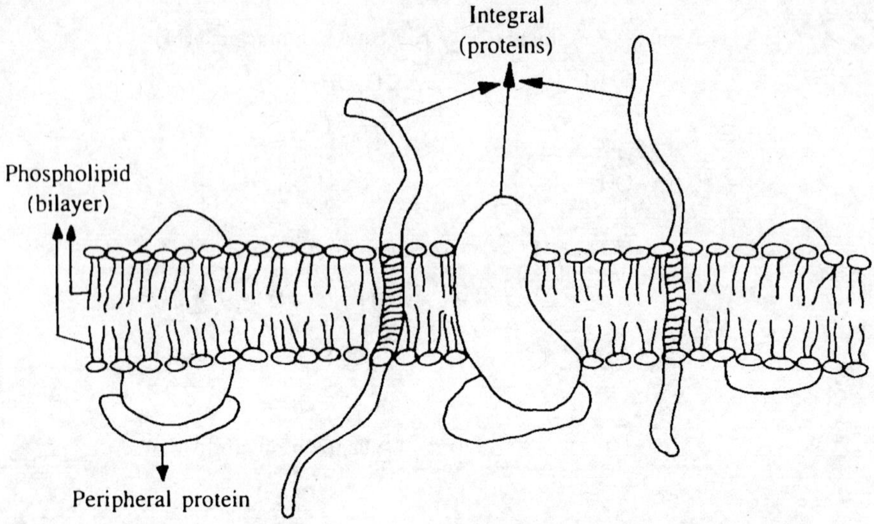

Fig. 4 Diagram showing plasma membrane (unit membrane).

Its exact function is not known. There are single-layered internal membranes in many autotrophic bacteria. In purple bacteria and cyanobacteria, they harbour photosynthetic pigments. These are the **thylakoid membranes**. Similar membranes are also present in nitrifying bacteria and are believed to aid in respiration.

The plasma membrane of eukaryotes is similar to that of eubacteria except that sterols are present in the membranes of eukaryotes, while they are absent in both eubacteria and archaebacteria. The membrane in archaebacteria has a different structure and composition from that of eubacteria. It is unilayered and is made up of branched chain lipids, linked together with **ether linkages** (in contrast to ester linkages of eubacteria). They are known chemically as **glycerol diethers** and **diglycerol tetraethers**. Non-polar lipids, known as **isoprenoids** are present in the membrane.

The plasma membrane functions as a selectively permeable membrane regulating the flow of materials into and out of the cell. Water passes through easily, as it is a small-molecule fluid without electrical charges. Non-polar lipids also pass through as they get dissolved in the membrane. Other substances like sugars, amino acids and inorganic ions are moved across the membrane by active mechanisms so that they can accumulate in the cell at concentrations higher than in the surrounding medium (against the concentration gradient).

Cytoplasm

In prokaryotes, the internal structure is uniform having a homogeneous protoplasm consisting of **cytoplasmic matrix** and filled with **ribosomes**. The nuclear material is distributed in the middle of the cell, and is known as **nucleoid**. It consists of a single DNA molecule lying in a highly coiled state. This is the bacterial chromosome. Besides the chromosome, many bacteria also possess small circular DNA molecules outside the main chromosome. They are the **plasmids**. Genes coding for reproduction, antibiotic resistance etc., are often located in the plasmids.

The ribosomes of prokaryotes are described as 70 S type, based on their sedimentation rate during centrifugation (Swedberg units). This is an important feature which distinguishes prokaryotes from eukaryotes (the latter have 80 S ribosomes). Each prokaryotic ribosome is made up of 50 S and 30 S subunits. The ribosomes contain protein and nucleic acid and are the sites of protein synthesis.

Granules of organic and inorganic storage substances are present in the cytoplasm. They are called **inclusion bodies**. The principal storage material in bacteria is **glycogen**, a branched polymer of glucose. **Poly-β-hydroxybutyrate** is another storage carbohydrate found in bacteria. Sulphur granules are formed in some sulphur oxidizing bacteria. **Carboxysomes** (containing the enzyme ribulose-1, 5 - carboxylase) are present in autotrophic bacteria. Many cyanobacteria form **gas vacuoles.** They consist of minute vesicles bound by a membrane and filled with gas. They give buoyancy to the cell so that it floats at or near the water surface (where light is available). Granules of polyphosphate, known as **volutin granules** are found in many bacteria, and they serve as a reserve source of phosphorus. They are also called **Metachromatin granules** as they appear reddish when stained with blue dyes such as methylene blue.

Endospores

Highly resistant spores are produced in some bacterial genera (especially *Clostridium* and *Bacillus*). They are resistant to heat and certain chemicals such as detergents. They are also resistant to desiccation and to radiation. They are called **endospores**, as they are formed inside the cell wall, which is retained. The cell is then known as a **sporangium**, because the endospore is enclosed inside. The spore may be located in the centre of the cell or towards a side. Sometimes the sporangium shows a bulge at the place of the spore (Fig. 5).

The endospore is highly refractile and special procedures are needed to stain it. Electron microscope study shows the endospore wall to be thick and complex, with many layers. The outermost layer, the **exosporium** is followed by the spore coat and cortex. The central core contains the cell wall, membrane and the cell contents. The wall contains **dipicolinic acid**. Endospores are important survival structures because of their highly resistant nature. They are usually formed under adverse conditions and can remain dormant for many years. They germinate when favourable conditions return. The spore wall and spore coat disintegrate by **lysozyme** that is produced, and a new vegetative cell emerges.

CYTOPLASM OF EUKARYOTES

The internal structure of a eukaryotic cell presents a complex picture. A

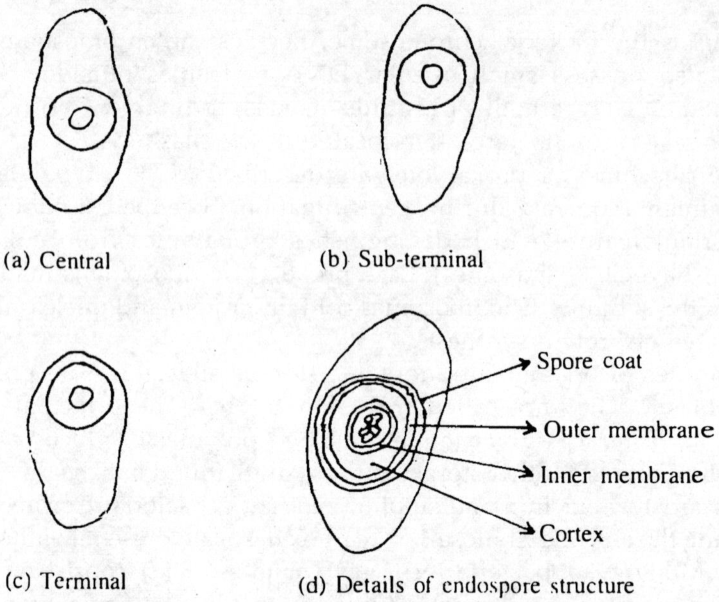

Fig. 5 Endospore bearing bacterial cells (sporangia) showing different positions of endospores.

distinct nucleus and subcellular organelles (mitochondria, plastids and golgi bodies) are present, and the cytoplasm is traversed with internal membranes. Cell division (described later) is a much more elaborate process in eukaryotes than in prokaryotes.

The cytoplasm consists of a *cytoplasmic matrix*, water amounting to 70–85% and the rest made up of proteinaceous microfilaments and microtubules, organized as a **cytoskeleton**. The matrix is also permeated by a network of flattened vesicles known as a **endoplasmic reticulum**. The **ribosomes** are either distributed in the cytoplasmic matrix or crowded on the endoplasmic reticulum. The endoplasmic reticulum is believed to aid in transport of proteins, lipids and other materials throughout the cell. The ribosomes are of the 80 S type (which distinguishes them from prokaryotic ribosomes which are of the 70 S type). Each ribosome is a dimer, made up of two 60 S and 40 S subunits, and has a molecular weight of about 4 million daltons.

The nucleus of eukaryotes is a spherical body enclosed in a double membrane known as **nuclear envelope**. It is continuous with endoplasmic reticulum. The outer and inner layers of the membrane are connected through **pores** which serve as channels of transport. The matrix of the nucleus inside the nuclear envelope is known as the **nucleoplasm**. It is traversed by a fibrous material known as **chromatin**. During cell division, chromatin is organized into individual chromosomes. Eukaryotes have more than one chromosome and each chromosome is made up of a DNA molecule, covered with a protein, called **histone**. A prominent internal structure of a nucleus is **nucleolus**. It generally disappears during cell division but reappears later.

Mitochondria are present in all eukaryotes. They are membrane-bound, with a smooth outer layer and a much-folded inner layer. The folds are known as **cristae**. Respiratory enzymes and electron transport carrier molecules are present in the mitochondria. Ribosomes are also present. Mitochondria possess small, circular DNA molecules which have a possible role in mitochondrial division during cell division.

Plastids are present in eukaryotic algae and higher plants. **Chloroplasts** are the most important of them, as they contain chlorophyll and are sites of photosynthesis. Chloroplasts are bound by a double-layered membrane. The matrix is known as the **stroma**. It consists of a DNA molecule, ribosomes, starch granules and lipid droplets. It is traversed by membranous sack-like structures known as **thylakoids**. At some places these thylakoids are packed in several layers and are known as **grana**. The presence of DNA in mitochondria and plastids indicates their independent ancestry. It is believed that they were endosymbionts in the cells and, in course of evolution lost their independent prokaryotic nature, becoming integral parts of eukaryotic cells.

The other prominent organelle present in many cells is **Golgi apparatus.** It is made up of a stack of closely arranged membranous sacks. The outer ends of the sacks cut off small vesicles from their edges. The sacks are called **dictyosomes.** They may secrete digestive enzymes or waste products, depending on requirement. Waste products are transported across the plasma membrane through these vesicles and excreted outside the cell.

External Features

Prokaryotes. Bacterial cell surface is usally covered by a slimy material known as **glycocalyx.** In some forms it is thicker in consistency and forms a well-marked envelope. Such a layer is called a **capsule** and is formed by a network of polysaccharide molecules extending from the surface of the cell wall. In a few bacteria it is formed from polypeptides (e.g. *Bacillus anthracis*). They contain large quantities of water and protect the cell from desiccation. They also aid in attachment of cells to solid substrates in water.

Many Gram-negative bacteria show very fine, short appendages extending outside and upto 100 nm in length and 3-10 nm in diameter. They are called **fimbriae** and are made up of microtubes of protein. They are believed to aid in the attachment of bacteria to solid surfaces. Somewhat longer appendages, usually 2–10 in number are present and these are known as **pili** (singular-pilus). They are known to aid bacterial conjugation and hence are also referred to as **'sex pili'**. The bacteriophages often use the pili as receptive sites for entry into the cell.

Several bacteria show **flagella** which aid in movement. The bacterial flagella are fine, thread-like structures, extending outside through the plasma membrane and cell wall. The place of attachment of the flagellum is known as the **basal body**. A hook extends from it and the flagellum extends to about 15–20 μm from the hook. Flagella are made up of a bundle of

3-4 protein molecules and the protein is called **flagellin.** The hook is made up of different proteins. The basal body has outer and inner ring-like structures connecting it to the plasma membrane and cell wall layers (Fig. 6). The flagella show a rotatory movement (acting like the oars of a boat) to propel the bacterium.

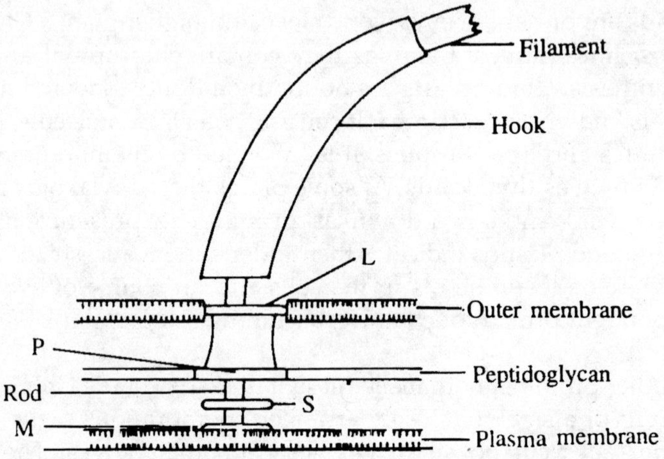

Fig. 6 Attachment of flagellum (basal body and hook) in *E. coli.*

Flagella are classified into the following types based on their arrangement (Fig. 7).

- Monotrichous: Single flagellum from a pole (e.g. *Pseudomonas*)

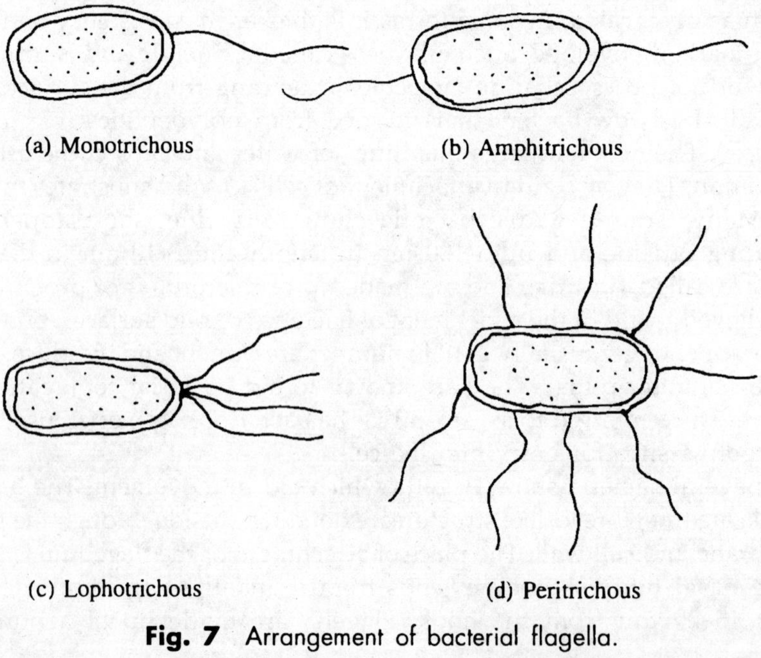

Fig. 7 Arrangement of bacterial flagella.

- Amphitrichous: One flagellum from each of the two poles (e.g. *Spirillum*)
- Lophotrichous: A tuft of flagella from a single pole (e.g. *Chromatium*)
- Peritrichous: Flagella arising from around the cell surface. (e.g. *Bacillus*)

In eukaryotes, some unicellular forms (e.g. *Chlamydomonas*) and also reproductive cells of some algae and fungi (zoospores), possess **flagella.** These are much more elaborate and much larger than bacterial flagella. They are from 100–200 μm in length and show undulatory movement. In a cross-section, they appear to be membrane-bound cylindrical structures, 0.2 μm in diameter, with a protein matrix. There are nine sets of double tubules inside the membrane and one set of two tubules at the centre (Fig. 8). The pattern is known as the 9+2 pattern of microtubules. Some cells show a large number of much shorter structures (5–20 μm) known as **Cilia**. They also have the same basic structure as flagella. The flagellum or cilium extends from a basal body in the cytoplasm. The flagella which have a smooth surface are said to be the **whiplash type**, while those that show numerous short lateral extensions are known as the **tinsel type**. The whiplash flagellum propels the cell forward from behind, while the tinsel type moves ahead, dragging the cell behind it. The cilia which occur around the cell, act like oars propelling the cell.

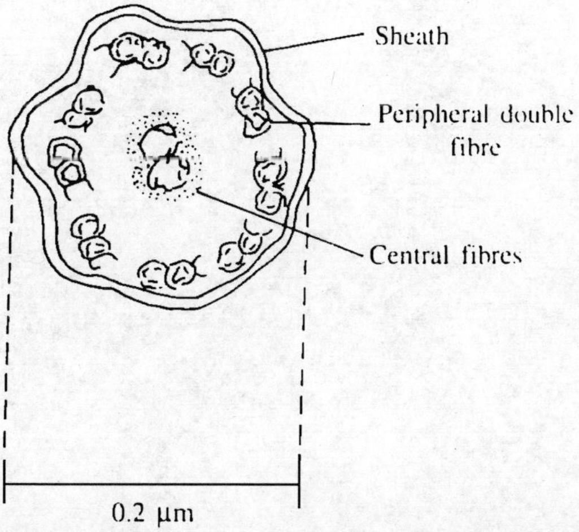

Fig. 8 Cross-sectional view of a eukaryotic flagellum showing microtubules.

CELL DIVISION

Cell division is the basis for increase in population of unicellular organisms and for growth and reproduction in multicellular organisms. It is a relatively simpler process in prokaryotes than in eukaryotes. In most bacteria and cyanobacteria, reproduction is by binary fission. The cell constituents increase quantitatively. The DNA gets attached to a part of the membrane and replicates. A wall is laid down separating the cell contents in such a way that a complete complement of macromolecules, monomers, inorganic ions and nuclear materials is present in each daughter cell. Cell division can be completed in 20 minutes to two hours in most bacteria. In some bacteria, multiplication takes place by **budding**. A bulge starts and enlarges from one end of the cell as the nuclear material is duplicated, and one of the complements of the nuclear material as well as other cell components enter the bud before the bud is partitioned from the parent cell by a cross wall.

In eukaryotic organisms, cell division in vegetative cells is by **mitosis.** The process occurs in such a way that individual chromosomes become distinct and duplicate. They gather at the centre of the cell. One set moves towards one pole and the other set to the opposite pole. A cell plate is then formed dividing the cell in two. During the formation of reproductive cells, the division is by **meiosis** in which chromosomes are arranged in homologous pairs at the centre of the cell. One chromosome of each pair moves to one pole and the other to the opposite pole. The result is that the number of chromosomes is reduced to half in each daughter cell. When fertilization occurs a male and a female cell unite, forming a **zygote** in which the original chromosome number is restored (Detailed processes of mitosis and meiosis are not elaborated here).

2

A SURVEY OF MICROORGANISMS AND VIRUSES

The cell is regarded as the basic unit of living organisms as it can carry on all life functions independently. The cellular organisms are further grouped broadly into *prokaryotes* and *eukaryotes*. The organization of the cell in prokaryotes is much simpler, and size of the cells is much smaller, than the highly complex and generally larger cells of eukaryotes. The prokaryotes and some groups of eukaryotes are considered as microorganisms. *The term microorganism is generally applied to any form of life that requires microscopic examination to recognize it.*

In addition to the microorganisms with cellular structure, there are a large number of forms which exhibit essential properties of life. *Viruses* and related forms belong to this category. If **self-perpetuation** is accepted as the most important criterion of life, the range of life forms encompasses self-perpetuating simple macromolecules and viruses at one end and the most complex and large plants and animals at the other. Of the microorganisms, bacteria are prokaryotes while algae, fungi and protozoa are eukaryotes. In recent years, fundamental differences were noticed in different bacteria and now they are separated into two groups, namely *archaebacteria* and *eubacteria*.

PRIONS

The simplest living entities discovered so far are *prions*. They are made up of only a small protein but are capable of infection and replication. They modify a host protein to replicate the prion molecule. The prions cause a few diseases in animals and man. *Scrapie* disease in sheep and *kuru* disease in humans are examples.

VIROIDS

Viroids are infectious agents consisting of a single-stranded RNA molecule. They lack a protein coat (which is characteristic of viruses) and hence do not show a definite shape. Coconut *Cadang-cadang* and *citrus Exocortis* diseases are examples of diseases caused by them. The RNA molecule is in the range of 240–375 nucleotides and shows a circular configuration. The viroid molecule is known to replicate in the host nucleus, utilizing the host RNA polymerase.

VIRUSES

Viruses are infectious agents, made up of protein and nucleic acid. The arrangement of the molecules imparts definite shape to the virus particles when they are outside the host cell. On entering a host cell, they behave as obligate parasites and multiply. They differ from cellular organisms in not having any metabolism and possessing only one type of nucleic acid, either DNA or RNA. The individual virus particle is called a virion. The protein molecules form an **outer coat**, enclosing the nucleic acid, which forms the **inner core**. The virions of viruses show a great range of variation in size, shape and symmetry of arrangement of molecules. Some viruses have a **lipid membrane** surrounding the virion. Between bacteria, which are cellular organisms and viruses which are particulate, is a group of pathogens with incipient cell structure. This includes *Rickettsias* and *Mycoplasmas*.

RICKETTSIAS

These are spherical or rod-shaped cells occurring as intracellular parasites. They possess a cell wall and a plasma membrane. They give Gram-negative reaction. The cells are small, 0.3 to 0.5 μm in diameter and 0.8 to 2.0 μm in length. They possess both DNA and RNA. The cells also divide by binary fission, as in bacteria. They are intracellular parasites and replicate in the host cytoplasm. Once their number increases, the host cell bursts, releasing them. *Rocky mountain spotted fever* and *Q-fever* are examples of diseases caused in humans by Rickettsias. They are transmitted by **arthropod vectors** (ticks etc.).

There is a monogeneric order, Chlamydiales, the genus being *Chlamydia*. It resembles Rickettsiales but has a unique feature of reproducing only within the cytoplasmic vesicles of the host cells. The cell of *Chlamydia* is Gram-negative, 0.2 to 1.5 μm in size and is an obligate parasite. *Lymphogranuloma venereum* is a sexually transmitted disease caused by *C. trachomatis*.

MYCOPLASMAS

Mycoplasmas are small, cellular organisms, which lack cell walls. The protoplasm is enclosed in a membrane. They are said to be **pleomorphic** as their shapes are variable due to the absence of a wall. Many of them have been cultured in the laboratory, on complex media, usually containing sterols. They are sensitive to tetracycline antibiotics. In culture they occur as **coccoid** (spherical) forms, 0.1 to 0.3 μm, and the colony appears as a 'fried egg', with a depressed centre and flat edges. They possess both DNA and RNA and are regarded as the smallest free-living cells. There is a related group, *Spiroplasmas* which have the same properties but the cells are spiral in shape. Examples of diseases caused by mycoplasmas are *chronic respiratory disease* of chickens and *pleuropneumonia* in cattle. Mycoplasma-like organisms also cause diseases in plants. It has, however, not been possible to culture them so far. Little leaf of brinjal, spike disease of sandal etc. are such diseases.

The basic features of different types of parasitic agents discussed above point to a gradually increasing complexity of organization, from single molecules to very small cells. Figure 9 gives the range of sizes of these groups.

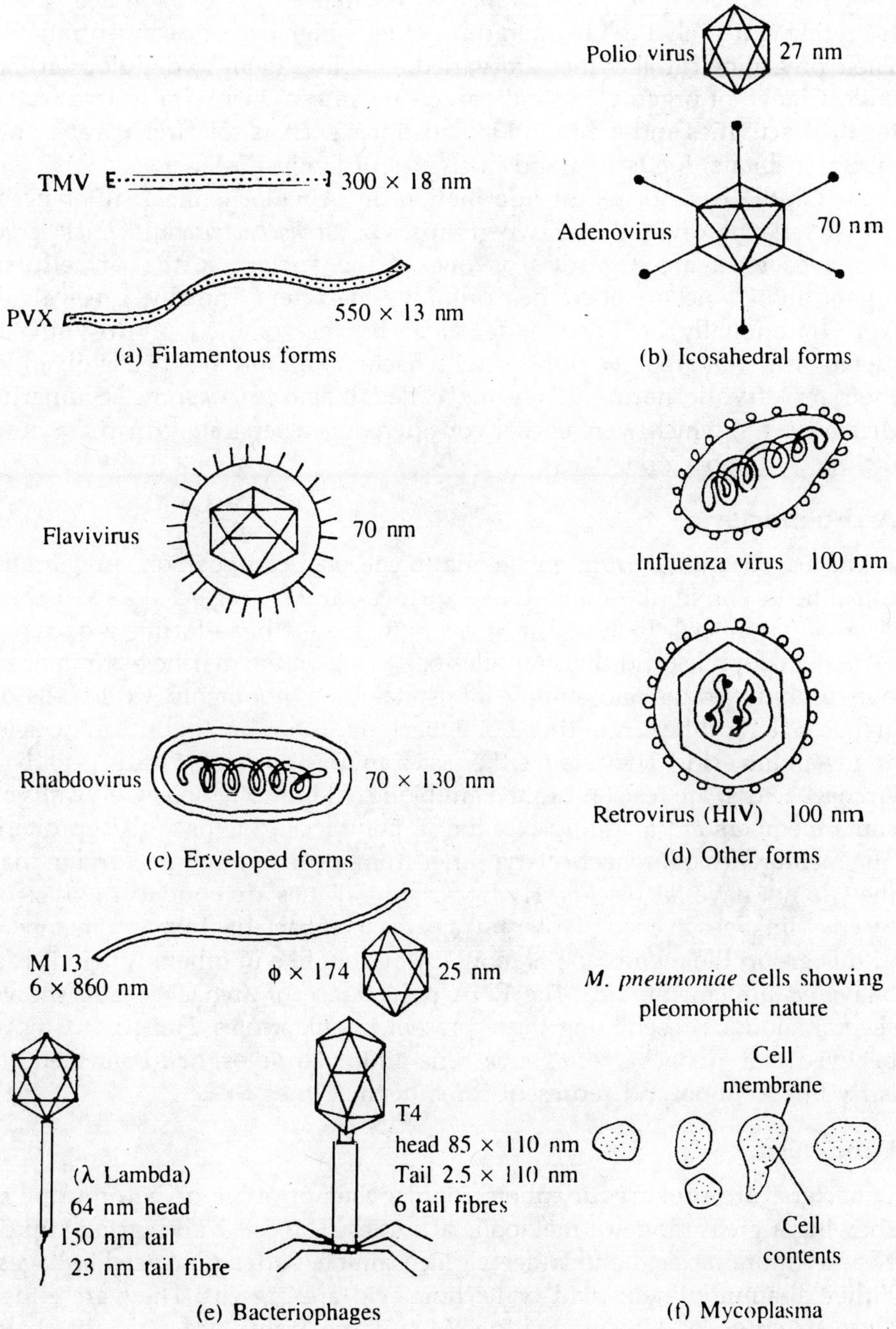

Fig. 9 Representative virus particles and mycoplasma.

BACTERIA

Bacteria represent the largest group of prokaryotes. They are essentially unicellular but may occur as groups of cells or show rudimentary mycelia. They are found in the widest range of habitats in nature, some of them tolerating extremely high temperatures or very high osmotic concentrations. Their physiological activities are varied, enabling them to develop on the widest range of organic and inorganic substrates. They require free water for their activities and are found in substrates such as soil, fresh water and marine habitats, foodstuffs and in tissues of living hosts.

On the basis of recent information of their biochemical differences, the bacteria are divided into two groups, viz. *Archaebacteria* and *Eubacteria*. Archaebacteria are regarded as one of the ancient forms of cellular organization which retained their primitive characters. The blue-green algae were traditionally included under algae because of their photosynthetic metabolism, but are now clubbed with bacteria, importance being given to their prokaryotic nature. They are called *Cyanobacteria* now. Similarly, *Actinomycetes*, which were earlier considered as a separate group are now included in eubacteria.

Archaebacteria

Archaebacteria differ from eubacteria in cell wall composition, membrane constituents and in ribosomal RNA sequences, and are regarded as a distinct class of organisms. Included in this group are methane-forming bacteria, extreme halophiles and thermophiles occurring on the marine bottom or in hot sulphur springs and similar inhospitable environments. Cell walls of archaebacteria differ from those of eubacteria in not having muramic acid and D-amino acids. They stain either as Gram-positive or as Gram-negative. Archaebacteria are resistant to the antibiotic chloramphenicol. Cell walls of some members are made up of either a polysaccharide or a glycoprotein. The membranes of archaebacteria differ from those of other bacteria in that they do not have fatty acids. Hydrocarbon moieties are bonded to glycerol by ether linkages instead of ester linkages as in eubacteria. In many members of this group the membrane is monolayered, unlike in others where it is a bilayered unit membrane. The RNA polymerase of archaebacteria shows several subunits resembling those present in eukaryotes. Hence, it is now believed that archaebacteria, eubacteria and eukaryotes, had branched off early in evolution and represent three distinct lines.

Eubacteria

Eubacteria are prokaryotic, phototrophic, chemotrophic or heterotrophic, showing a great range of metabolic activities. The cell wall is made up of N-acetylmuramic acid and N-acetylglucosamine, with amino acid linkages. Either di-aminopimelic acid or teichoic acid are present. They are either Gram-positive or Gram-negative. Membrane lipids are straight-chain molecules, connected with ester linkages, resulting in a bilayered structure. The RNA polymerase is made up of four subunits. Eubacteria are sensitive to chloramphenicol.

CYANOBACTERIA

Cyanobacteria are prokaryotes but resemble algae in that they possess chlorophyll-*a* and show oxygen evolving photosynthesis, which is absent in eubacteria. They have **phycobiliproteins** which act as accessory pigments in photosynthesis. The photosynthetic pigments are bound to membranes known as **thylakoids**. Cyanobacteria are essentially unicellular but in many genera the cells are enclosed in mucilaginous sheaths, forming filaments. Some of the cells are modified, lack chlorophyll and are known as **heterocysts**. Nitrogen fixation occurs in those cells.

PROCHLOROPHYTA

This recently discovered group has attracted considerable attention because of the resemblance of its photosynthetic pigments with eukaryotic algae and higher plants on the one hand, and cyanobacteria on the other. Oxygen is evolved in photosynthesis, as in the above. Two genera, *Prochloron* and *Prochlorothrix* are known. The former is unicellular and occurs as a symbiont with marine invertebrates. As in cyanobacteria, chlorophyll-*a* is present but phycobiliproteins are absent. Instead, chlorophyll-*b* is present as in algae and plants. The cell is prokaryotic and the photosynthetic system consists of **thylakoid membranes**. The second genus, prochlorothrix occurs in shallow lakes as a phototroph and could be cultured in the laboratory.

PROTOZOA

Protozoa are eukaryotic, unicellular organisms traditionally studied by zoologists but are now placed in eukaryotic protista. The cells are larger and possess a distinct nucleus and subcellular organelles. Cell walls are absent and the protoplast is surrounded by a membrane. Some members have a protective covering known as the **testa**. It is made up of calcium carbonate. Under adverse conditions many of them develop a protective wall resulting in the formation of cysts. The protozoan cell is larger than the prokaryote cell. Protozoa lack chlorophyll and are heterotrophic. They require water for movement. Most of them live in soil and water but some are parasites on animals and man. African sleeping sickness caused by *Trypanosoma gambiense*, amoebic dysentery caused by *Entamoeba histolytica* and malaria caused by *Plasmodium* spp. are examples of serious diseases of humans caused by protozoan parasites. In the aquatic environment, protozoa form a significant component of plankton.

Free-living protozoa obtain their food by ingesting particles such as organic debris, bacterial cells etc., by a process known as **pinocytosis**. Amoeboid forms have flexible shape and the membrane extends and surrounds the particle. The extensions are known as **pseudopodia**. In others, a cavity is present in the cell and particles are drawn into it. Some protozoa have fine cilia extending from the surface of the cell membrane, while some others possess flagella. Protozoa are capable of motility either by amoeboid movement, or through the action of cilia or flagella. Reproduction is by binary fission or conjugation. Figure 10 illustrates the four main orders into which protozoa are classified.

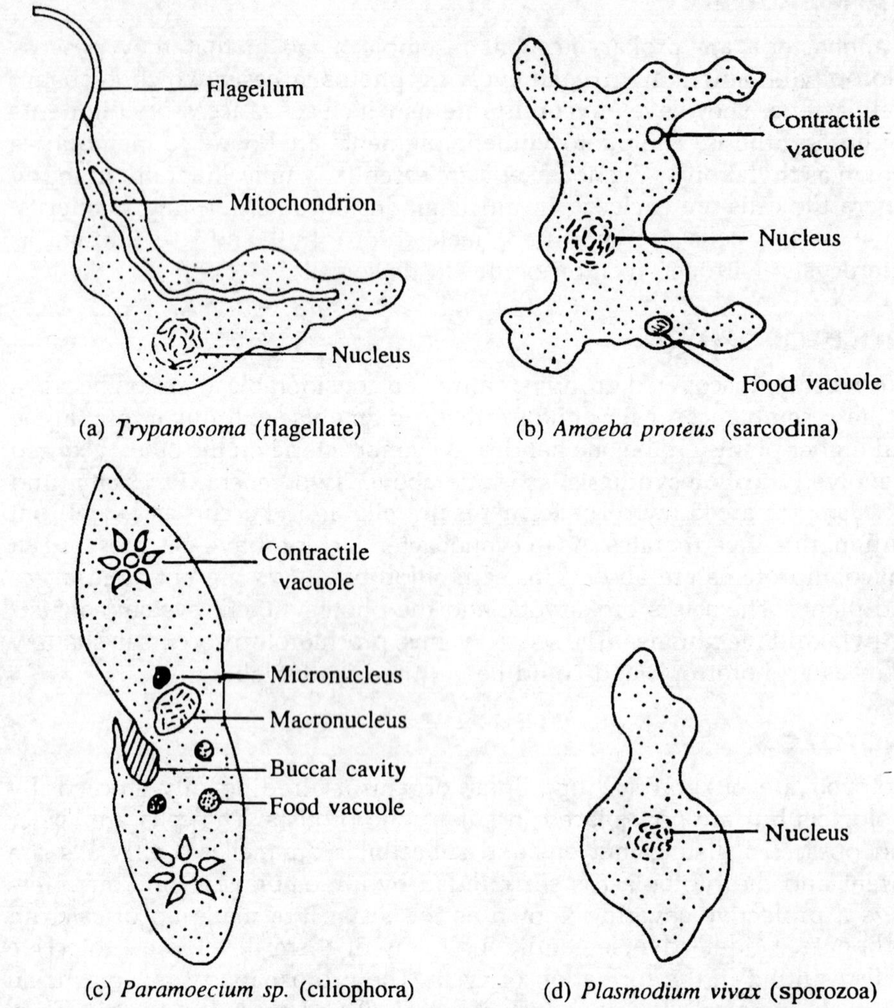

(a) *Trypanosoma* (flagellate) (b) *Amoeba proteus* (sarcodina)

(c) *Paramoecium sp.* (ciliophora) (d) *Plasmodium vivax* (sporozoa) trophozoic stage

Fig. 10 Representative protozoan forms.

Mastigophora (Flagellates)

These are free-living in fresh water or parasites on animals and man. Locomotion is by flagella, e.g. *Trypanosoma* (it has a long, twisted mitochondrion).

Sarcodina (Amoebae)

The members of this order are naked in vegetative phase (**amoeboid**) or enclosed in a shell (**Foraminifera**). Free-living in fresh waters or marine habitats or as parasites. Some inhabit mouth parts and intestines, e.g. *Amoeba proteus.*

Ciliophora (Ciliates)

All members of this order possess cilia around the cell during at least a part of their life cycle. Largest of protozoan forms. The cell is elaborate showing an oral cavity (mouth and a connected gullet). Food particles are drawn into the gullet where they get digested. There are two nuclei, a larger macronucleus and a smaller micronucleus. The latter has been shown to determine inheritance, e.g. *Paramoecium.*

Sporozoa (Sporozoans)

All members are obligate parasites on man, animals and birds. They are nonmotile in the vegetative stage, produce sporelike bodies (sporozoites), through which they are transmitted. Some have alternate hosts. The malarial parasite *Plasmodium* is a sporozoan.

ALGAE

A very large group of eukaryotic organisms, algae occur in fresh water, marine habitats and in soil. They are the principal producers of organic carbon and represent the first trophic level in aquatic environments. They carry out oxygenic photosynthesis. They are diverse, ranging from unicellular forms to large and complex thalloid structures and their reproductive methods are also varied. Like the other eukaryotes, they possess a distinct nucleus and sub-cellular organelles. Cell walls are present in the vegetative state. Many unicellular forms show motility by flagella. Many forms produce motile zoospores and gametes in reproduction. Photosynthetic pigments occur in plastids which are made up of thylakoids enclosed in a membrane. The thylakoids occur in parallel bundles separated by a matrix. These bundles are referred to as **grana** and the intervening matter as stroma. Chlorophyll pigments are attached to the thylakoids. Chlorophylls *a, b, c, d,* and *e* are present in different groups of algae. Carotenoids and xanthophylls are also present. In one group, (*Rhodophyta*) phycobiliproteins also occur. Mitochondria are present in the cytoplasm. Algae are divided into a number of classes, based on structure, pigment systems and methods of reproduction. Table 3 gives the principal features of each class and Fig. 11 illustrates some representative forms.

FUNGI AND SLIME MOULDS

Fungi are eukaryotes in which chlorophyll is absent. They show a great range in thallus organization, from **unicellular** to **coenocytic filaments** (lacking crosswalls) and septate filaments. The filaments are made up of branched hyphae forming the thallus, referred to as **mycelium**. They show both asexual and sexual methods of reproduction. Higher fungi produce fruit bodies following sexual reproduction. The fruit bodies are sometimes large and complex and are made up of thickly woven hyphae, forming **pseudoparenchymatous** tissue. Asexual reproduction is mostly by spores. Fungi occur as **saprophytes** on a wide variety of substrates in soil, water and spoiled food materials, or as **parasites** on plants and animals. They cause highly destructive diseases on crop plants. One of the best known

Table 3 Important Features of Main Groups of Algae

Algal class	*Common name*	*Morphological range*	*Cell wall constituents*	*Pigments*	*Storage products*	*Habitats*	*Examples*
Chlorophyta	Green algae	Unicellular, colonial, filamentous, heterotrichous or coenocytic forms	Cellulose	Chlorophylls *a* and *b*	Starch sucrose	Mainly fresh water. A few occur in soil and marine habitats	*Chlamydomonas* (unicellular) *Cladophora* (filamentous) *Fritschiella* (Heterotrichous) *Codaeum* (coenocytic) *Volvox* (colonial)
Euglenophyta	Euglenoids	Unicellular	Absence of cell wall	Chlorophylls *a* and *b*	Paramylon	Fresh water	*Euglena* (flagellate)
Chrysophyta	Golden brown algae and diatoms	Unicellular	Silica	Chlorophylls *a, c* and *e*	Lipids	Fresh water, marine and soil	*Navicula* (diatom)
Phaeophyta	Brown algae	Filamentous, leafy and some are very large and plantlike	Cellulose	Chlorophylls *a* and *c* xanthophylls	Laminarin, mannitol	Marine	*Padina* (leafy) *Fucus* (large and plant-like)
Pyrrophyta	Dinoflagellates	Unicellular with several flagella	Cellulose	Chlorophylls *a* and *c*	Starch	Fresh water and marine	*Gymnodinium*
Rhodophyta	Red algae	Unicellular, filamentous, leafy	Cellulose	Chlorophylls *a* and *d* phycocyanin and phycoerythrin	Floridean starch, Fluoridoside	Marine	*Polysiphonia* (filamentous)

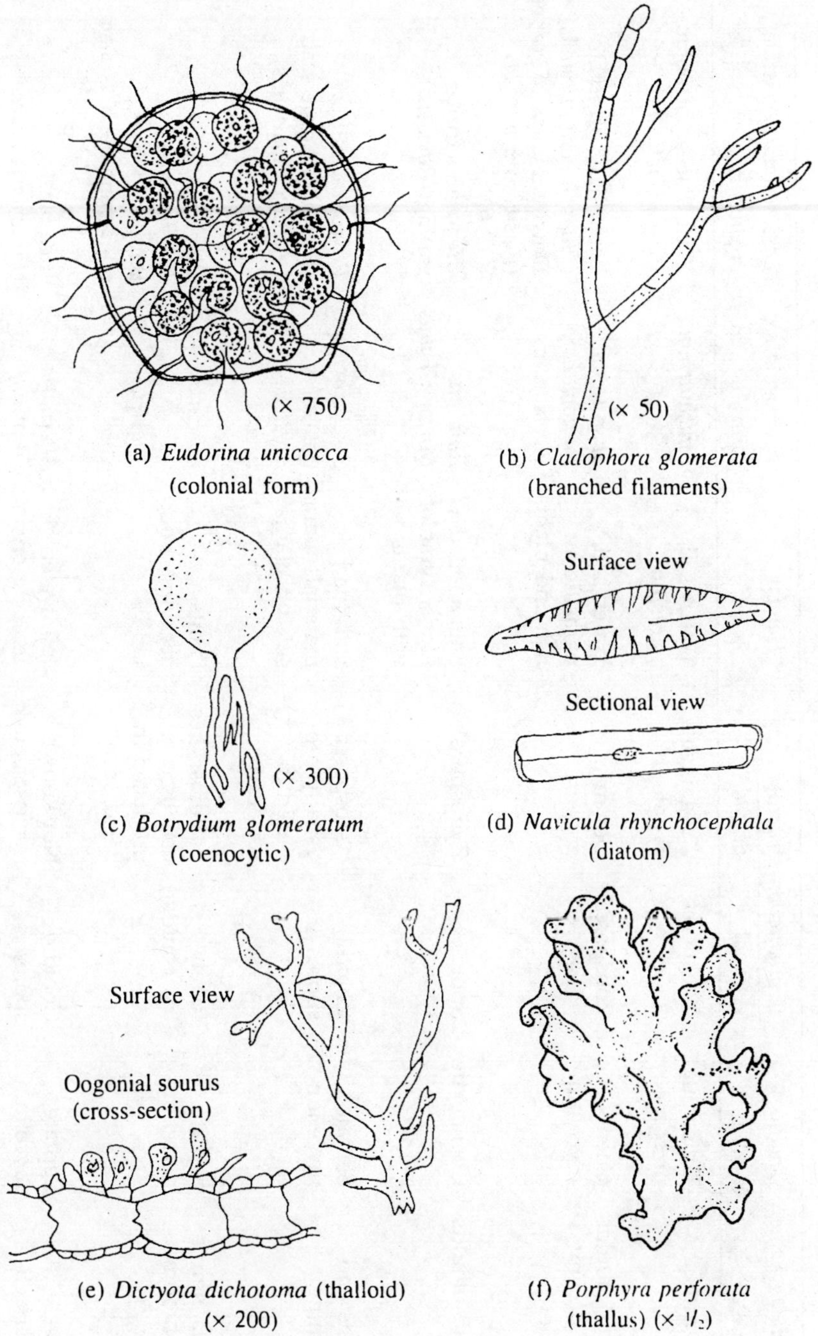

Fig. 11 Representative forms of algae.

fungi is **yeast**, which has been extensively studied for its biochemistry and industrial uses. The principal features of different classes of fungi are listed in Table 4 and representative forms are illustrated in **Fig. 12.** A few fungi produce *mycorrhiza* occurring symbiotically in roots of higher plants.

Table 4 Important Features of Main Groups of Fungi

Fungal class	*Common name*	*Thallus*	*Reproduction structure*		*Habitat*	*Representative forms*	*Diseases and causative organisms*
			Asexual	*Sexual*			
Chytridiomycetes	Chytrids	Unicellular or coenocytic	Zoospores	Zoogametes, zygote	Water, soil, parasitic	*Synchytrium*	Wart disease of potato (*S. endobioticum*)
Oomycetes	Water moulds and downy mildews	Coenocytic mycelium	Zoospores	Oogamy, oospore	Water, soil, parasitic on fish and plants	*Saprolegnia* *Phytophthora* *Plasmopara*	Late blight of potato (*Phytophthora infestans*) Downy mildew of grapes (*Plasmopara viticola*)
Zygomycetes	Bread moulds	Coenocytic	Sporangio-spores	Zygospore	Soil, spoiled foods, symbiotic with plants	*Rhizopus*, *Mucor Glomus*	Bread and fruit rotting (*Rhizopus*)
Ascomycetes	Yeasts and sac fungi	Septate mycelium	Conidio-spores	Ascospores usually produced in Ascocarps	Soil, food materials, parasitic on plants	*Saccharomyces* (yeast) *Erysiphe* *Peziza*	Powdery mildew (*Erysiphe*) Apple scab (*Venturia inaequalis*)
Basidiomycetes	Rusts, smuts mushrooms	Septate mycelium	Rarely (by conidia)	Basidiospores usually produced in Basidiocarps	Plants, decaying wood, soil	*Puccinia* *Agaricus*	Black rust of wheat (*P. graminis*) Smut of wheat, sugarcane etc. (*Ustilago*)
Deuteromycetes	Fungi imperfecti	Septate mycelium	Conidio-spores	Not known (or associated with asco- or basidiomycetes)	Soil, water, decaying organic matter, parasites on plants	*Alternaria* *Fusarium* *Aspergillus* *Penicillium*	Rice blast (*Pyricularia oryzae*)

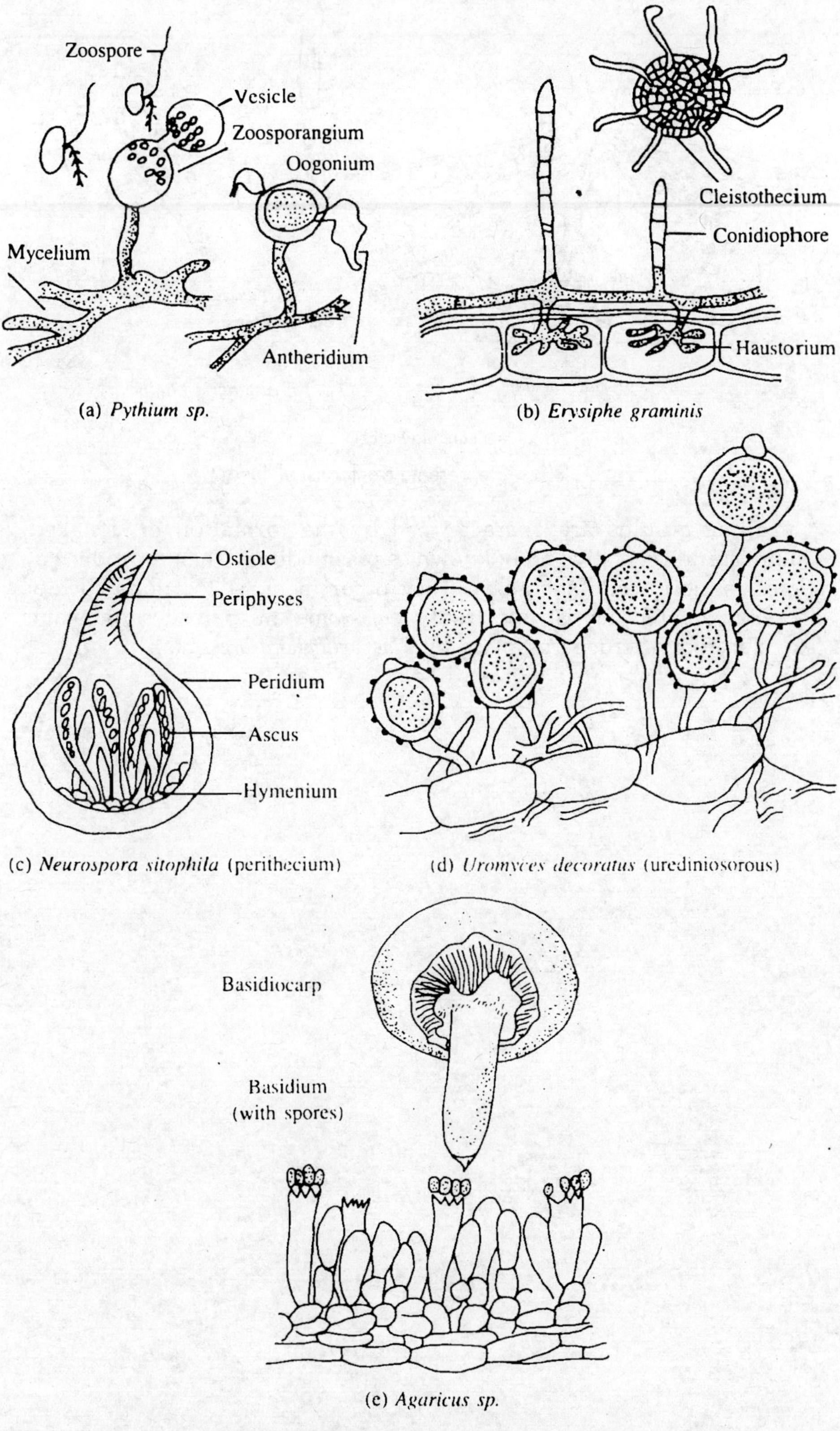

Fig. 12(a) Representative forms of fungi.

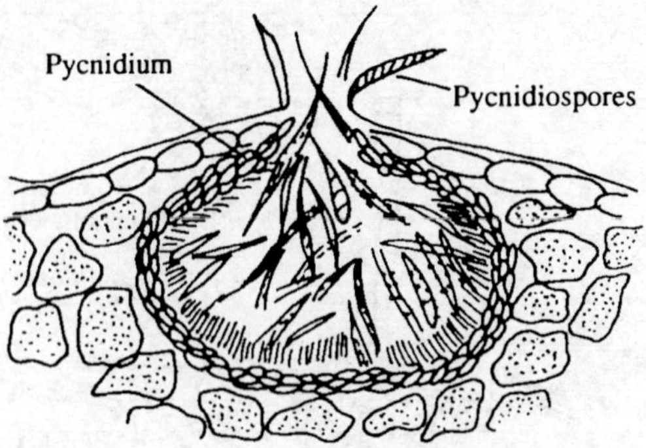

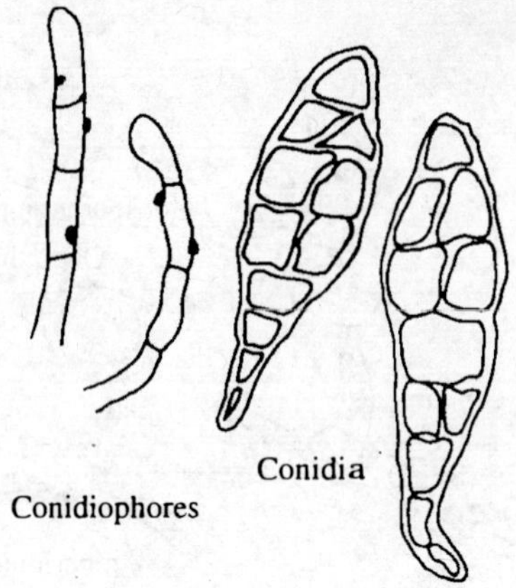

(deuteromycetes)

Fig. 12(b) Representative forms of fungi.

Slime moulds are characterized by the formation of a naked, multinucleate vegetative body known as **plasmodium.** They reproduce by spores formed inside sporangia. They occur in moist locations such as decaying organic debris, tree trunks etc. Some are parasites on plants (e.g. club root of crucifers caused by *Plasmodiophora brassicae*).

3

MORPHOLOGY AND CLASSIFICATION OF BACTERIA

The extreme physiological diversity of microorganisms and their simple morphology, coupled with rapidly advancing knowledge have made their classification an extremely difficult task. Conventional classification takes into account various features of morphology, nutrition, pigment, sensitivity to antibiotics, habitat requirements, pathogenicity, symbiosis and immunological relations. Numerical taxonomy was evolved, based on the above characters, taking the aid of computer analysis to determine the relations. This has been further improved by taking up molecular approaches, especially DNA base composition, nucleic acid hybridization, nucleic acid sequences in RNA etc. The classification system is undergoing continuous change. In a recent publication by Woese, et. al. (1990) all cellular organisms are divided into three groups: **Archaea, Bacteria** and **Eukarya.**

MORPHOLOGICAL FEATURES

On the basis of shape, bacteria are grouped into four types: the generic name usually indicates shape and arrangement of cells.

Cocci (Singular-coccus): Spherical cells, occurring either singly or in groups, features useful in identification. Usually small (0.5 to 1.5 µm in diameter).

- **Monococcus** (Single cells)
- **Diplococcus** (Occurring in pairs of cells)
- **Streptococcus** (Chains, formed due to cell division in one plane and cells not separating)
- **Staphylococcus** (Cells dividing in random planes and adhering together) and
- **Sarcina** (Cells dividing in three regular planes forming cubical groups).

Bacilli (Singular-bacillus): Rod-shaped cells, usually occurring singly, sometimes in chains, generally 1.0 to 6.0 µm in length and 0.5 to 1.5 µm in diameter. The shape shows variation with the ends being flat, rounded or cigar-shaped (e.g. *Bacillus, Clostridium*).

Vibrios. Rod-shaped cells with distinct curvature appearing like a comma, or as a twisted rod (e.g. *Vibrio cholerae*).

Spiral. Long and slender cells twisted into spirals or helical shapes. They are called **spirilla** if rigid, and **spirochaetes** if flexible. Many members are 2.0 to 4.0 μm long and 0.25 μm thick, but some are very long, reaching 250 μm in length.

Other shapes. Actinomycetes develop long filaments forming rudimentary or well developed **mycelia.** *Corynebacteria* have cells with irregular shapes, appearing to be modified rod-shaped forms. *Cyanobacteria* develop large cells (up to 7.0 μm in diameter), which may occur singly (unicellular) or in unbranched or branched filaments, called *Trichomes.* The mycoplasmas lack a cell wall and show variation in shape. Such forms are said to be **pleomorphic.**

(The external appendages of cells, i.e., flagella, fimbriae, pili and capsules have already been described.)

COLONY CHARACTERS

The appearence of bacterial colonies when cultured on semisolid media are also useful for identification. When inoculated on to a solid medium, bacteria develop into colonies of characteristic size, shape, colour, sheen, elevation and margin, which aid in identification. Common colony features are listed below and are illustrated in Fig. 13.

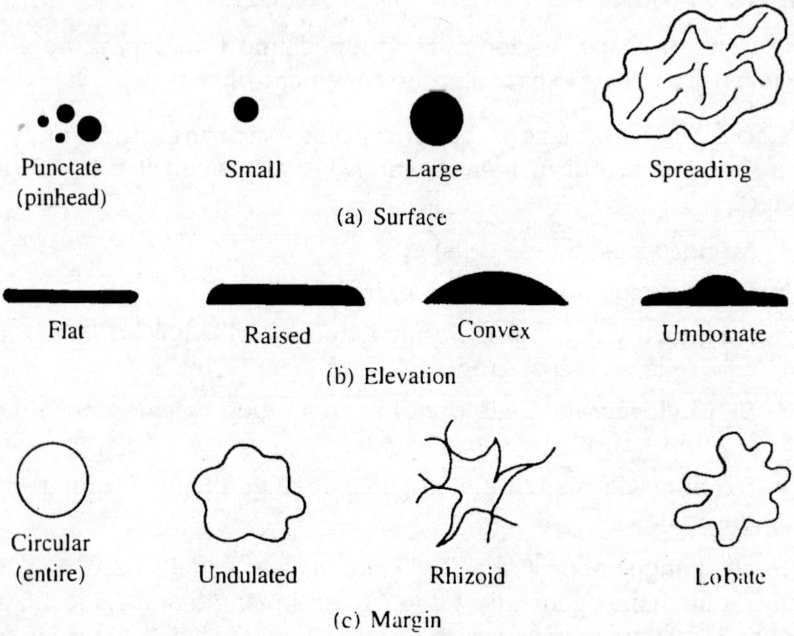

Fig. 13 Morphological features of bacterial colonies.

- **Colour** Translucent, white, creamy, yellow, red.
- **Elevation** Flat, raised, convex, umbonate.
- **Sheen** Smooth and shining, dull.
- **Margin** Circular, undulated, rhizoid, lobate.
- **Size** Punctate (pinhead colonies), small, large, spreading.

PHYSIOLOGICAL AND GENETIC CHARACTERS

Biochemical and metabolic features are very useful in bacterial identification. Features such as utilization of various carbon and nitrogen sources, cell wall constituents, motility of cells, oxygen relations, growth factor requirements, cardinal temperatures and pH range for growth, photosynthetic pigments, osmotic relations, formation of secondary metabolites, sensitivity to metabolic inhibitors and antibiotics, fermentation products, pathogenic and symbiotic relations etc., are employed for this purpose. Molecular characteristics include comparison of proteins, enzyme properties and DNA base composition. The latter has proved to be very useful as different organisms show a characteristic range of G + C content (guanine + cytosine). The melting temperature of the DNA molecule, at which it breaks up into individual nucleotide bases gives a measure of G + C percentage. The G + C of many bacteria has been determined and has proved useful for comparing relationships.

Genetic factors such as conjugation pattern, transformation between species, and presence or absence of plasmids, have been used in classification. Nucleic acid hybridization study indicates the degree of closeness of different organisms, based on the extent of hybridization between the DNAs of test organisms. More recently, techniques for determining sequences of bases in nucleic acid molecules have been developed and have proved to be very valuable in determining relationships.

Thus, a number of parameters are studied when an organism has to be identified or when comparisons with other organisms are made.

NUMERICAL TAXONOMY

A system of computer analysis of data obtained on selected parameters has been evolved for grouping the organisms into taxonomic units. The coefficient of association is calculated between each pair of characters studied, and the coefficients are arranged to form a **similarity matrix.** Organisms showing great similarity are grouped together and separated from dissimilar organisms.

MAJOR GROUPS OF BACTERIA

As the knowledge on bacteria is accumulating rapidly, their classification is undergoing frequent changes and no permanency can be assigned. However, major characteristics form the basis for classification. A recent edition of Bergey's *Manual of Systematic Bacteriology* has divided bacteria into 33 groups. The characteristics of major groups are summarized below. Important members of the groups are also discussed. Cyanobacteria and archaebacteria are discussed separately. Table 5 lists major groups of bacteria, with examples.

Table 5 Genera of Bacteria Mentioned in the Text

Section	*Genera mentioned*
A: Gram-negative eubacteria:	
Spirochaetes	*Treponema, Leptospira*
Aerobic, Gram-negative rods and cocci	*Pseudomonas, Azotobacter, Rhizobium, Bradyrhizobium, Agrobacterium*
Methylotrophs	*Methylomonas, Methylococcus*
Enterobacteriaceae	*Escherichia, Salmonella, Shigella Yersinia, Citrobacter, Klebsiella, Proteus*
Vibrionaceae	*Vibrio, Photobacterium*
Sulphur-reducing bacteria	*Desulfovibrio, Desulfomonas*
Spirillaceae	*Azospirillum, Aquaspirillum, Bdellovibrio*
Other Gram-negative forms	*Rickettsia, Coxiella, Chlamydia, Mycoplasma*
B: Gram-positive eubacteria:	
Gram-positive cocci	*Micrococcus, Staphylococcus, Streptococcus, Enterococcus, Lactococcus, Leuconostoc*
Endospore forming rods and cocci	*Bacillus, Clostridium*
Non-sporing rods	*Lactobacillus, Arthrobacter, Corynebacterium*
Mycobacteria	*Mycobacterium*

Table 5 Genera of Bacteria Mentioned in the Text (Cont.)

Section	*Genera mentioned*
C: Photosynthetic bacteria and cyanobacteria:	
Purple sulphur bacteria	*Chromatium, Thiospirillum*
Purple non-sulphur bacteria	*Rhodopseudomonas, Rhodospirillum*
Green sulphur bacteria	*Chlorobium*
Green non-sulphur bacteria	*Chloroflexus*
Cyanobacteria	*Anabaena, Nostoc, Spirulina, Lyngbya, Gloeocapsa, Stigonema*
D: Aerobic chemolithotrophic bacteria:	
Nitrifying bacteria	*Nitrosomonas, Nitrosococcus, Nitrosospira, Nitrobacter*
Sulphur oxidizing bacteria	*Thiobacillus, Thiomicrospora.*
Gliding bacteria	*Cytophaga, Sporocytophaga, Leucothrix, Thiothrix, Beggiatoa*
E: Actinomycetes	
Nocardioforms	*Nocardia, Faenia*
Forms with multilocular sporangia	*Frankia, Geodermatophilum*
Actinoplanetes	*Actinoplanes, Micromonospora*
Streptomyces and related forms	*Streptomyces*
Maduromycetes	*Streptosporangium*
Thermomonospora and related forms	*Thermomonospora, Thermoactinomyces*
F: Archaebacteria:	
Methanogenic forms	*Methanobacterium, Methanococcus, Methanosarcina, Methanospirillum, Methanopyrus*
Extreme halophiles	*Halobacterium*
Extreme thermophiles	*Thermoproteus, Sulfolobus*
Cell wall-less form	*Thermoplasma*

Gram-negative bacteria. **Spirochaetes** are characterized by long and slender cells, with a flexible, helical shape, 0.1 to 3.0 μm × 5 to 250 μm in size. Because of the helical structure, they show motility in liquids even though they lack flagella. The plasma membrane, cell wall and outer membrane resemble those of other Gram-negative forms. Spirochaetes are chemoheterotrophs and may be aerobes/anaerobes (facultative or obligate) occurring on a wide range of substrates. Some are pathogens. *Treponema pallidum* causes syphilis. *Leptospira* is found in the soil.

Aerobic Gram-negative rods and cocci. A number of genera are included in this group. *Pseudomonas* has about 90 species, occurring on diverse habitats such as soil, food materials and mouth parts of man and animals. Some are pathogenic on crop plants. The cells are straight or slightly curved rods, with one or more polar flagella and 0.5 to 1.0 μm × 1.5 to 5.0 μm in size. *P. aeruginosa* is a widespread species and has been used in many physiological studies. *P. fluorescence* produces a greenish fluorescent pigment.

Azotobacter is a free-living nitrogen-fixing bacterium. The cells are large cocci, 1.5 to 2.0 μm in diameter, occurring in soil and water.

Rhizobium is a rod-shaped bacterium, 0.5 to 0.9 μm × 1.2 to 3.0 μm in size which fixes nitrogen symbiotically in leguminous plants. The cells assume pleomorphic forms in root nodules and are known as **bacteroids** in that stage. There are both fast and slow growing species in media. The latter are now separated into a new genus, *Bradyrhizobium*. A genus related to *Rhizobium* is *Agrobacterium*. It is a common soil organism. *A. tumefaciens* forms tumours in many plants. The tumour-inducing principle (TI plasmid) is now used in genetic engineering as a vector in gene transfer.

Methylotrophs. These are methane-oxidizing bacteria. There are two genera, *Methylococcus* (spherical and non-motile) and *Methylomonas* (straight or curved rods with a single polar flagellum). They oxidize methane or methanol and are called **methylotrophs**. They are aerobic or microaerophilic and grow above anaerobic locations in soil or water, where methane is formed.

Enterobacteriaceae. The members of this family are Gram-negative, straight rods, non-motile or motile with peritrichous flagella, occurring in intestines, and on other substrates also. Some of them cause gastrointestinal or urinary tract infections. *Erwinia* causes soft rots in crop plants. *Salmonella* and *Shigella* cause typhoid and dysentery. *Escherichia coli* is the common colon bacterium and the most thoroughly studied organism on morphological (especially under electron microscope), physiological and genetic aspects. It is being used extensively in gene transfer studies and successfully for the production of drugs such as insulin and interferon. Plague is caused by *Yersinia*, a member of this group. *Citrobacter*, *Klebsiella* and *Proteus* are other enterobacteria. They ferment sugars producing gas. Fermentation products of different genera of this family are useful in their identification.

Vibrionaceae. The members of this group are straight or curved rods, Gram-negative and motile with polar flagella 0.3 to 1.3 × 1.0 to 3.5 μm in size. Most of them are aquatic organisms and show **bioluminiscence**, emitting blue-green light. *Photobacterium* lives symbiotically in luminescent organs of fish. *Vibrio cholerae* causes the dreaded cholera, while some other species cause fish diseases.

Sulphur-reducing bacteria. These are Gram-negative, morphologically diverse bacteria which are obligate anaerobes. They reduce sulphur, sulphate or other oxidized sulphur compounds, forming hydrogen sulphide. *Desulfovibrio* and *Desulfomonas* use only elemental sulphur. They occur in lake and marine sediments and waterlogged soils.

Spirillaceae. The members of this group are helical or vibrioid forms, Gram-negative, motile with polar flagella and are aerobic or microaerophilic. They are found in soil, fresh water and marine habitats. *Azospirillum* occurs in the rhizosphere of grasses and legumes and fixes nitrogen. *Aquaspirillum* occurs in fresh water and *Oceanospirillum* in the sea. The spirilla show a great range of sizes, from 0.2 to 1.7 × 2.1 to 60.0 μm. Spirilla differ from spirochaetes in that they are rigid (spirochaetes have flexible cells).

An interesting bacterium, related to spirilla is *Bdellovibrio.* It is a curved rod with a polar flagellum, 0.2 to 0.5 × 0.5 to 1.4 μm in size. It preys on other Gram-negative bacteria by penetrating them and growing between the cell wall and the plasma membrane. The host cell becomes round and the plasma membrane breaks, leaking the cell contents which are utilized by the *Bdellovibrio.* It grows into a long filament, which divides to form a number of cells. This is described as **multiple fission.** The host cell wall is lysed and the daughter cells are released. This predator bacterium is aptly named *B. bacteriovirus.*

Other Gram-negative forms. The *Rickettsiae, Chlamydiae* and the *Mycoplasmatales* form this group. The first two are intracellular parasites, living and reproducing inside host cells. *Rickettsia* and *Coxiella* are human pathogens. *Chlamydia trachomatis* affects humans and birds. The mycoplasmas are widespread in nature and some of them are pathogenic on animals and plants. *Mycoplasma mycoides* causes pleuropneumonia in humans.

Gram-positive eubacteria. The Gram-positive bacteria except mycobacteria and actinomycetes are included in this section.

Gram-positive cocci. About 18 genera are included in this group. *Micrococcus* and *Staphylococcus* are small, 0.5 to 2.5 μm in diameter, and divide in more than one plane, producing clusters of cells. Colonies of micrococci are often coloured (yellow, orange or red). They occur in soil and water and also on skin of mammals. They are aerobic and catalase-positive. Species of *Staphylococcus* are facultative anaerobes and usually form irregular clusters and ferment glucose anaerobically. They have **teichoic**

acids in the cell wall. They inhabit the skin. *S. aureus* is pathogenic and causes boils and abscesses in humans.

The genus *Streptococcus* is a facultative anaerobe, catalase-negative and ferments sugars without gas formation, the mechanism being homolactic fermentation. About 29 species are recognized. They cause haemolysis of erythrocytes in the blood. Pyogenic streptococci are associated with pus formation. *S. pyogenes* causes sore throat etc. In recent years, the genus has been divided into three genera, *Streptococcus, Enterococcus* and *Lactococcus* on the basis of DNA hybridization patterns. The genus *Leuconostoc* carries out heterolactic fermentation. The above four genera as well as the rod-shaped *Lactobacillus* are commonly called **lactic acid bacteria**.

Endospore-forming rods and cocci. Six genera of endospore-forming bacteria are included in this section. The genus *Bacillus* is aerobic or microaerophilic while *Clostridium* is an obligate anaerobe. Some species have peritrichous flagella. Both are widespread in soil but also occur in other habitats. Due to the presence of heat-resistant endospores they survive drying. Bacillus species produce antibiotics like **bacitracin** and **polymyxin**. *Bacillus anthracis* causes anthrax in animals and humans. *Bacillus cereus* causes food poisoning. *Bacillus thuringenesis* forms a protein crystal which has insecticidal properties. *Clostridium* causes spoilage of canned foods. Several species produce toxins which cause food poisoning. *Clostridium tetani* is the cause of tetanus infection.

Non-spore forming rods. The genus *Lactobacillus* develops in milk and forms lactic acid as a fermentation product. The cells are bacilli or coccobacilli (short rods), facultative anaerobes or microaerophils, have complex nutritional requirements and lack catalase activity. They are found in milk and milk products, meat, fruits and in water. Both homolactic and heterolactic fermenters are present. They prefer slightly acidic conditions, around pH 4.5 to 6.4. Lactobacilli are used in making **curd (yogurt)** and other fermented foods and silage. *Lactobacillus bulgaricus* and *Streptococcus thermophilus* are the common milk fermenters. *Arthrobacter* is a common soil organism. It is aerobic and catalase-positive. Its reproduction is by what is described as **snapping division**.

Corynebacterium is characterized by straight or curved rods or club-shaped forms. It is an aerobe or facultative anaerobe, catalase-positive and reproduces by snapping division. Many species are saprophytes but some are plant or animal pathogens. *Corynebacterium diphtheriae* causes diphtheria in humans.

Mycobacteria. This group consists of only one genus, *Mycobacterium.* They are straight or slightly curved rods, sometimes branching to form filaments. They are aerobic and catalase-positive. They are said to be acid fast, i.e., when they are stained with basic fuschin, the stain cannot be washed out by acid alcohol. They share this property with some actinomycetes. The acid fast reaction is due to the presence of lipids outside the peptidoglycan layer. Mycobacteria are either saprophytes or pathogens on ruminant animals, primates and humans. Tuberculosis, which is a worldwide problem is caused by *M. tuberculosis.* As it can spread from cattle, milk is pasteurized or boiled to prevent infection from this source. Another serious disease is leprosy caused by *M. leprae.* The mycobacteria are slow-growing and difficult to eradicate.

Photosynthetic bacteria and Cyanobacteria. There are five groups of photosynthetic bacteria, which carry out **anoxigenic photosynthesis,** except Cyanobacteria which are oxigenic. Oxygen is not evolved in these forms while it is produced in Cyanobacteria. Three of the groups deposit sulphur inside or outside the cell. Four groups are photoautotrophic while one group is photoheterotrophic. The photosynthetic pigments are *bacterio-chlorophylls* in the four groups while in Cyanobacteria it is chlorophyll a (as in algae and green plants) with phycobiliproteins as accessory pigments. In Cyanobacteria, **purple sulphur bacteria** and **purple non-sulphur bacteria** photosynthetic pigments occur on **Lamellar membranes** known as **thylakoids**, which are connected to the plasma membrane. **In Green sulphur bacteria**, and **Green non-sulphur bacteria**, they occur in membranous vesicles, independent of the plasma membrane.

The Cyanobacteria are a large, diverse and widespread group. About 24 genera are included. They are unicellular or filamentous. The filaments are unbranched or branched collections of cells enclosed in common mucus. They are called **trichomes**. The cells range from 1.0 to 10.0 μm in diameter. Some members occur as colonies. Reproduction is by binary fission, or budding, or fragmentation of trichomes or colonies. Large, thick-walled cells known as **akinetes** are produced in soil. They survive adverse conditions and form new filaments on germination. *Gloeocapsa* is unicellular, *Lyngbya* is an unbranched filament and *Stigonema* has branched filaments. Many filamentous forms also have non-pigmented cells called **Heterocysts**. Nitrogen fixation is carried out in those cells. Because of efficient photosynthetic pigments and nitrogen-fixing ability, cyanobacteria are adapted to occur in a wide range of habitats where there is moisture such as water bodies, soil, including desert areas and locations of feeble light. They are an important source of nitrogen in rice fields (e.g. *Anabaena, Nostoc).* Filaments of *Spirulina* are used as food in some desert areas of Africa and it is now cultured as a source of single-cell protein.

Aerobic, Chemolithotrophic bacteria. The bacteria that derive their energy and electrons from reduced inorganic compounds are included in this group.

Nitrifying bacteria. At present nine genera are recognized. Of those, *Nitrosomonas,* which oxidizes ammonia to nitrite and *Nitrobacter,* which further converts nitrite to nitrate, are widespread in soil. They are Gram-negative, strict aerobes with high rates of respiration. They are about 0.6 to 1.0 × 1.0 to 2.0 μm in size and motile with polar flagella. *Nitrosococcus* (spherical), *Nitrosospira* (spiral) and *Nitrococcus* (spherical) are other important genera. *Nitrobacter* reproduces by budding and the other multiply by binary fission.

Sulphur oxidizing bacteria. There are two sub-groups in this. *Thiobacillus* and *Thiomicrospora* are Gram-negative rods, whereas *Beggiatoa*, *Thiothrix* etc. are filamentous, gliding bacteria. *Thiobacillus thiooxidans* and *T. novellus* oxidize hydrogen sulphide, sulphur or thiosulphate to form sulphate. *T. ferrooxidans* also uses ferrous iron. *T. denitrificans* produces nitrogen gas from nitrates thus causing loss of nitrate from soil. The sulphur oxidizers are important in soil fertility. Of the **gliding bacteria**, *Cytophaga* and *Sporocytophaga* are slender rods, capable of decomposing cellulose, the most common and resistant polysaccharide. *Beggiatoa* and *Leucothrix* form slime-covered trichomes and glide along substrates, leaving slime trails. *Beggiatoa* occurs in sulphur-rich habitats such as sulphur springs. Another group of gliding bacteria are *Myxobacteria*. They show a complex life cycle, producing an aggregation of cells into a fruit body. They are most common in soils (e.g. *Myxococcus*).

Actinomycetes. The members of this large and important group are mainly soil inhabitants. They were regarded as intermediate forms between bacteria and fungi on account of their mycelial structure, but are now grouped with eubacteria because of their prokaryotic nature and cell wall composition. When growing on agar media, they produce mycelium which is partly aerial and partly submerged in the medium, formed from much-branched but thin hyphae. Most members also produce single-celled spores. Thus they resemble fungal colonies in culture. In liquid media also, they form individual colonies and do not make the medium turbid like other bacteria. The spores are formed either externally as **conidia** or inside sporangia as **sporangiospores.** The hyphae usually divide into long cells with more than one **nucleoid** in them. The spores are easily detached and disbursed. Many actinomycetes produce pigments of various shades. The cell walls are chemically distinguishable into four types. While most actinomycetes are saprophytes, a few of them are pathogens also.

They are classified on the basis of mycelial and spore morphology, pigments, secondary metabolites and cell wall features. Representative genera are illustrated in Fig. 14.

Actinomycetes, particularly the genus *Streptomyces* are economically important. Majority of the clinically useful **antibiotics** are produced by actinomycetes (their industrial importance will be presented later). Taxonomically, the following six groups are recognized.

Nocardioforms. (e.g. *Nocardia, Faenia*). They develop substrate mycelium which easily breaks up into individual cells. They are mostly saprophytes in soil and water but *Nocardia asteroides* causes a lung disease in humans and animals of low resistance.

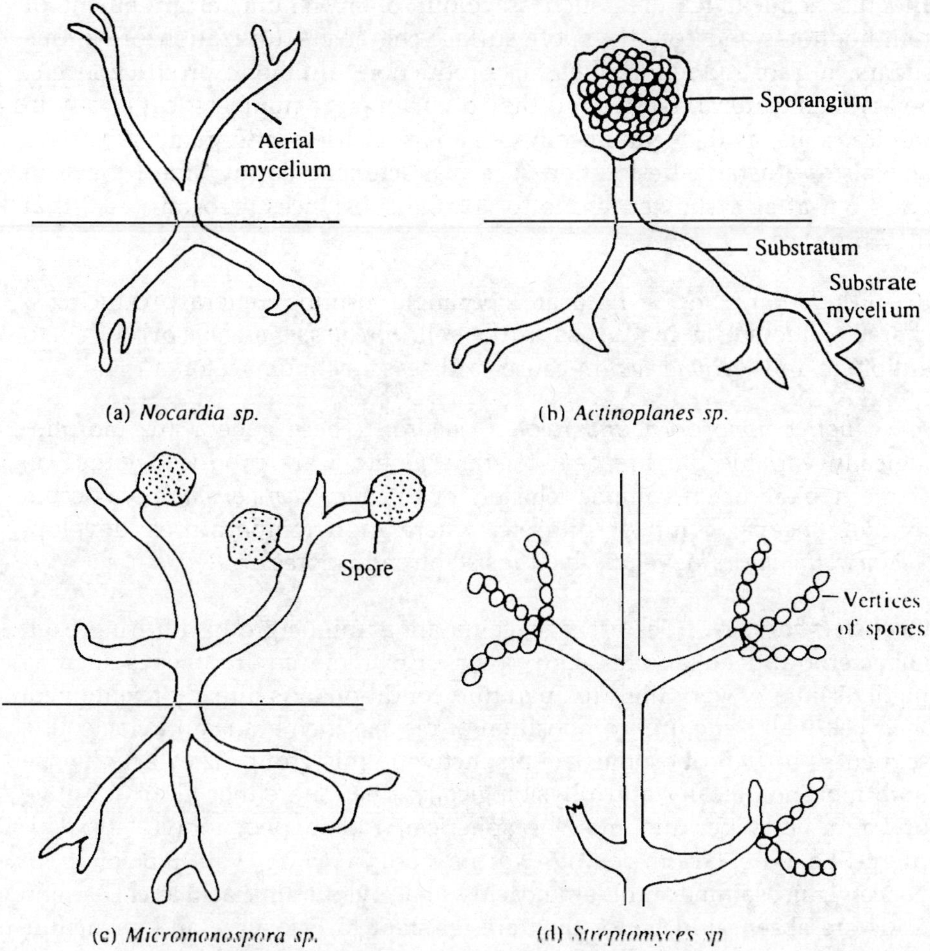

Fig. 14 Representative forms of actinomycetes.

Actinomycetes with multilocular sporangia. This section includes *Frankia* and *Geodermatophilus*. The hypha divides in two planes to produce clusters of spores. *Frankia* forms root nodules in non-leguminous trees, in symbiotic association. The spores of *Geodermatophilum* are motile.

Actinoplanetes. In this group, extensive substrate mycelia are formed but aerial mycelium is absent or rudimentary, with sporangiospores rising above the agar surface. The spores are arranged in coils or parallel rows in the sporangium (e.g. *Actinoplanes*). *Micromonospora* produces single sporangiospores on branched sporophores.

Streptomyces and related forms. *Streptomyces* comprises a large genus with nearly 380 species. They are aerobic, produce aerial and substrate mycelia and chains of nonmotile spores (conidia) enclosed in a thin fibrous sheath. Different species are distinguished on the basis of morphological and biochemical features such as colour of mycelium, arrangement of conidiophores and conidia, spore surface characters, utilization of various sugars, nitrate reduction, melanin production, antibiotic production etc. Soil is their natural habitat and they occur in large numbers, especially in tropical soils, as they can decompose a large variety of organic substrates and also withstand desiccation. The characteristic earthy smell that soil gives out after a shower is due to volatile substances (geosmin etc.) that they produce.

Maduromycetes. These are sporangia-forming genera producing a characteristic sugar, **madurose** in the cell. *Streptosporangium* produces an antibiotic and *Actionomadura* causes a disease, actinomycetoma.

Thermomonospora and related genera. These genera are morphologically variable but are heat tolerant. Conidia are produced either on single hyphae or on compact clusters of hyphae. *Thermomonospora* occurs in dung heaps, compost pits etc. where high temperatures develop. *Thermoactinomyces* develops heat-resistant endospores.

Archaebacteria. The archaebacteria are a unique group differing from eubacteria and eukaryotes, and occur in extreme environments such as highly saline or very high temperature conditions. As already mentioned, their cell wall structure, composition of plasma membrane and nucleotide sequences in 16 S ribosomes are distinctive to this group. They are diverse both morphologically and physiologically. They are either Gram-positive or Gram-negative, and may be spherical, rod-shaped, lobed, spiral or plate-like. The Gram-positive forms have a thick wall made up of N-acetylglucosamine and L-aminoacids. N-acetylmuramic acid and D-amino acids are absent and hence they are resistant to lysozyme and penicillin. The Gram-negative archaebacteria have a relatively thick protein or glycoprotein cell wall, but the outer membrane is absent.

The plasma membrane differs from that of eubacteria and eukaryotes in having branched chain hydrocarbons attached to glycerol, connected by ether links (not straight-chain fatty acids connected by ester links as in eubacteria). There are polar lipids, phospholipids, glycolipids etc. in the membrane. The archaebacterial chromosome is a single circular DNA molecule as in eubacteria but smaller in length (molecular weight of about 0.8 to 1.1×10^9 daltons, compared to 2.5×10^9 daltons of *E. coli*). The ribosomes and RNA polymerase of archaebacteria share some properties with eubacteria and some with eukaryotes.

Autotrophs and organotrophs are present among archaebacteria. The methane-forming methanogens and extreme thermophiles occur in unusual habitats like hot sulphur springs. The autotrophs fix CO_2 by a reverse TCA cycle or reductive acetyl CoA pathway, leading to the production of glucose, via pyruvate. Five groups of archaebacteria are recognized.

Methanogenic archaebacteria. These are strict anaerobes obtaining their energy by converting substrates such as CO_2, H_2, acetate, formate and methanol to methane (CH_4) or methane and CO_2. They are the methane producers in anaerobic digesters and *gobar* gas plants and hence, an important source of renewable energy. There are 13 genera which include long rods or filaments (*Methanobacterium*), straight or curved rods (*Methanosarcina*) and helical forms (*Methanospirillum*). Two are Gram-positive, *Methanosarcina* is Gram variable while the others are Gram-negative. In nature, they occur in anaerobic situations rich in organic matter such as swamps, lake or sea bottom sediments and rice fields. An extreme thermophile, *Methanopyrus kandleri* occurs in 'hot vents' in oceans, with a temperature range of 85 to 110°C, the optimum being 98°C. Although methane is an important and clean energy source, it is also one of the greenhouse gases contributing to global warming.

Extreme halophiles. These are aerobic and heterotrophic, growing on complex, proteinaceous substrates. *Halobacterium* grows in highly saline conditions (36% NaCl) and is dependent on NaCl for stability of the cell wall. Halobacteria occur in salt lakes and in packets of salted food. *Halobacterium salinarium* is unique in having a purple pigment, bacteriorhodopsin which facilitates photosynthesis without chlorophyll. Many halobacteria have yellow or red pigments which protect them from strong sunlight.

Extreme thermophiles metabolizing sulphur. These are extreme thermophiles and most of them are strict anaerobes using sulphur as electron acceptor. They grow in hot sulphur springs or soil rich in elemental sulphur. *Thermoproteus*, grows at 78 to 96°C, a strict anaerobe, either as a heterotroph on organic substrate, or as a lithotroph, fixing CO_2 by using H_2 or S or CO as electron donors. *Sulfolobus* grows in acidic conditions (pH 2 to 3), at temperatures of 70 to 80°, and is known as a **thermoacidophile**. It is a Gram-negative, lobed bacterium and is aerobic, oxidizing sulphur to sulphuric acid.

Cell wall-less archaebacteria. *Thermoplasma* grows in refuse piles of coal mines which contain iron pyrite (FeS) which is oxidized to sulphuric acid by lithotrophs. It results in highly acidic and hot conditions. *Thermoplasma* grows well at 55 to 59°C. It does not possess a cell wall but the plasma membrane is reinforced by diglycerol tetraethers. DNA is stabilized with a histone-like protein, thus resembling the chromosome of eukaryotes.

Sulphate reducers. These bacteria (e.g. *Archaeolobus*) reduce sulphate to form hydrogen sulphide. They are extremely thermophilic and strictly anaerobic.

Thus, archaebacteria are unusual organisms adopted to life in extreme environmental conditions.

4

VIRUSES

Viruses are infectious agents occurring on plants, animals, humans as well as microorganisms. They are in the form of particles outside but behave as obligate parasites on entering host cells. Certain diseases whose causative agents were earlier not known are now shown to be caused by viruses.

A virus particle is essentially made up of a single type of nucleic acid (either RNA or DNA) which is enclosed in a protein **coat.** The nucleic acid is called a **core** and the particle is known as a **virion.** The protein coat is called a **capsid** and along with the nucleic acid it is called a **nucleocapsid**. In some viruses the virion is enclosed in a **lipid** or **lipoprotein envelope**. Viruses do not possess any metabolism of their own and use the host metabolic and genetic apparatus for their multiplication. This process is called *replication,* in contrast to cell division in cellular organisms. The virus particles are extremely small, ranging from about 20 nm (nanometers) to 400 nm.

The viruses are classified and subdivided into groups on the basis of physicochemical properties of the virion and the nature of hosts on which they occur. For convenience, they are classified as **plant viruses, animal viruses** and bacterial viruses or **bacteriophages**.

All viruses have only one type of nucleic acid, possess one or more proteins, do not have any metabolic activity and reproduce by replication.

SYMMETRY OF VIRUSES

Besides the coat protein and nucleic acid core, some viruses have enzyme or other proteins inside the coat. The manner in which the coat proteins and the nucleic acid core are arranged determines the symmetry of virus particles. Three main types of symmetry are present as shown in Fig. 15.

Helical Symmetry

The nucleic acid core is arranged in a coiled spring-like manner and is enclosed in a protein coat in which the subunits of protein are also arranged in a helical manner. Such virions appear as rods or filaments and are referred to as **rod-shaped, tubular** or **filamentous** viruses, and their symmetry as **helical symmetry**.

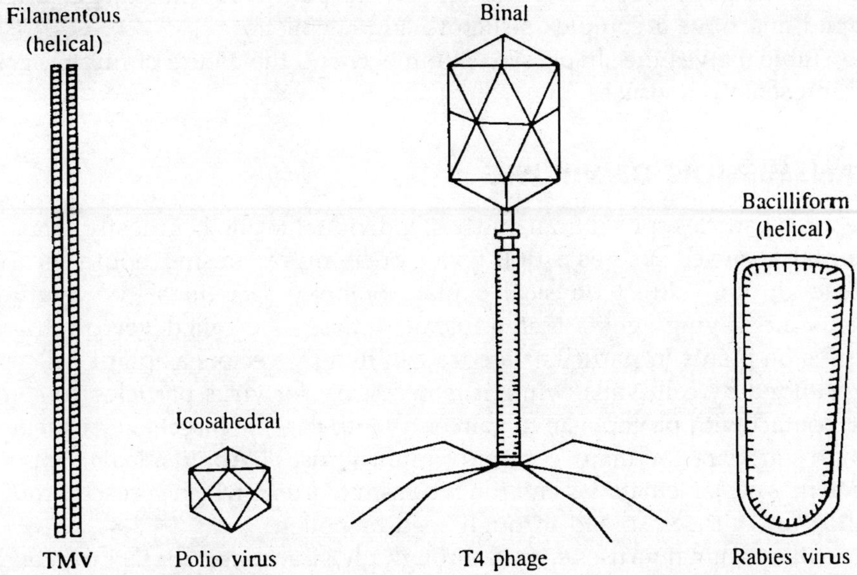

Fig. 15 Main types of symmetry in viruses.

Icosahedral Symmetry

Here the capsid forms a spherical structure with the protein subunits arranged in groups. The protein subunits occur in groups of five molecules (pentamers) or six molecules (hexamers). Each group is known as a **capsomer**. The arrangement of capsomers is such that a crystal-like structure, with 20 triangular faces and 12 vertices (points where adjacent triangles meet) is formed. This is an *icosahedron* and the symmetry is known as icosahedral symmetry. The nucleic acid occurs as a tightly coiled molecule in the space inside the capsomers.

Binal Symmetry

A few bacteriophages have a complex structure with the particles having a combination of icosahedral and helical symmetries. There is an icosahedral head to which a helical tail is attached. Such an arrangement is known as **binal symmetry.** The nucleic acid core is enclosed in the head. The tail attaches itself to the host cell. Accessory proteins in the form of connecting plates and tail fibres also occur in some cases.

Other Forms

Some viruses are cylindrical or **bullet-shaped** particles. They are usually **enveloped**. The **lipid** or **lipoprotein** envelope is derived from host plasma membrane or synthesized by the host as the particle leaves the cell. The envelopes of some viruses show protein or glycoprotein projections known as **spikes**. The nucleic acid is helically arranged in these bullet-shaped viruses. Some viruses are **pleomorphic**, having flexible shapes. They also show helical symmetry. The viruses of the **pox group** are large, brick-shaped and have a complex structure and chemistry.

Table 6 gives the shape, size, symmetry and the nature of nucleic acid in representative viruses.

TRANSMISSION OF VIRUSES

The virus particles spread from one host individual to another mostly through external agencies such as wind, water or living organisms. Some viruses spread through direct physical contact or body secretions like sputum, mucus etc. Living agents that transmit diseases are called **vectors.** Most viruses on plants in particular are transmitted by vectors as plant cells are surrounded by cell walls, while it is necessary for virus particles to come into contact with protoplasm to gain entry into the cell. Insects with sucking mouth parts are the main vectors of plant viruses. (Aphids, leaf hoppers, white flies etc.). Nematodes and fungi transmit viruses which spread through soil. A few viruses spread through seed or pollen.

Many animal viruses spread through physical contact or through body fluids. Water is also an important source of transmission (polio virus etc.). Since animal cells are covered only by plasma membrane the virus particles can directly enter the cells. The membrane surrounds the particle which is drawn inside. This process is known as **pinocytosis**. Some animal viruses are transmitted by vectors (mosquitoes etc.).

The bacteriophages also have to penetrate cell walls to gain entry into the host protoplasm. The particle gets attached to a **receptive site** on the wall or on cellular projections such as **pili.** The nucleic acid of the phage is then injected or drawn inside, leaving the protein coat outside. Figure 16 illustrates the entry of viruses into host cells.

COMPOSITION OF VIRUSES

The nucleic acid component of the virus, either RNA or DNA constitutes from about 3 to 30% of molecular weight of the virion. It is generally lower in helical viruses (3–6%) and higher in icosahedral ones (10–30%).

The nucleic acid is the genome of the virus and directs the synthesis of all components. The nucleic acid may be a single-stranded or a double-stranded molecule. Most plant viruses have single-stranded RNA genomes (ssRNA) while animal viruses and bacteriophages show all four types i.e. ssRNA, dsRNA, ssDNA or dsDNA. The single-stranded genome (ssRNA or ssDNA) may directly function as mRNA, or as a template for mRNA synthesis in which case it is known as **positive strand** (or "+" strand). If it acts as a template for the formation of a complimentary strand, which then forms mRNA, it is known as a **negative strand** (or "–" strand). The nucleic acid molecule may be an open-ended molecule in which case it is said to be **linear,** or close-ended when it is called a **circular** molecule. When a single-stranded genome forms a complimentary strand, it becomes

Table 6 Some Properties of Bacteriophages and Viruses (Representative Forms)

Virus (Name and Group)	*Principal host*	*Viral genome*		*Symmetry*	*Virion size*	*Special features*	*Remarks*
		Nucleic acid	*Molecular wt or nucleotide bases*				
Bacteriophages:							
MS 2	*E. coli*	ssRNA "+"	3569 bases	Icosahedral	26 nm	180 protein subunits	Genome acts as mRNA Overlapping of one gene occurs
$\phi \times 174$	*E. coli*	ssDNA "+"	5386 bases	Icosahedral	25 nm	Spikes present on vertices	Replicative form produced Overlapping of genes occurs
M13	*E. coli*	ssDNA "+"	7200 bases	Helical (filamentous)	6 x 860 nm	Released without host cell lysis	Genome enters cytoplasm while coat protein is retained in plasma membrane; useful as cloning vector
T7	*E. coli* (Shigella etc.)	dsDNA	39936 base pairs	Binal (tadpole-like)	Head: 75 nm Tail: 20 nm	Infects only female cells through pili	Bidirectional DNA replication. 25 genes identified
T4	*E. coli*	dsDNA	Mol. wt. 120×10^6d	Binal (tadpole-like)	Head: 110×85 nm Tail: 110×25 nm	Head is pronate icosahedron. Base plate and tail fibres	DNA has 5-hydroxymethyl cytosine
λ (lambda)	*E. coli*	dsDNA	48503 base pairs	Binal (tadpole-like)	Head: 65 nm Tail: 150 nm Tail fibre: 23 nm	Lysogeneic	DNA integrates with host genome in lysogenic state. Circularization occurs in free state
$\phi 6$	*Pseudomonas*	dsRNA	Mol. wt. of segments: 2.3, 3.1 & 5.0×10^6d	Icosahedral	75 nm	Enveloped	3 segments of RNA present (divided genome)
Animal Viruses:							
Polio virus (Picornaviridae)	Humans	ssRNA "+"	Mol. wt. 2.6×10^6 d	Icosahedral	28 nm	3 antigetic types Immunity permanent	Genome acts as mRNA coding for a polyprotein
Rabies virus (Rhabdoviridae)	Canines and Humans	ssRNA "–"	Mol. wt. 3.5×10^6 d	Bullet-shaped (helical)	70×175 nm	Enveloped	Forms replicative form in host

Table 6 (Cont.)

Virus (Name and Group)	Principal host	Viral genome		Symmetry	Virion size	Special features	Remarks
		Nucleic acid	Molecular wt/ or nucleotide bases				
Influenza virus (Orthomyxoviridae)	Humans	ssRNA "–"	Mol. wt. 4.0×10^6 d	Pleomorphic (helical)	80 × 120 nm	Enveloped with haemaglutinin and neuraminadase	3 antigenic types. Segmented genome
Reovirus (Reoviridae)	Humans	dsRNA	Mol. wt. 1.3×10^6 d	Icosahedral	75 nm	2 protein coats present	Segmented genome. Each transcribes a ssRNA which acts as mRNA
Parvovirus (Parvoviridae)	Mammals	ssDNA "–"	4800 bases	Icosahedral	25 nm		Smallest genome in animal viruses. Overlapping of genes
ARD virus 4 (Adenoviridae)	Humans and mammals	dsDNA	Mol. wt. 20×10^6 d	Icosahedral	80 nm	Extended spikes from the vertices	Virion assembled in host nucleus
SV 40 (Papovaviridae)	Mammals	dsDNA	5243 base pairs	Icosahedral	45 nm	Transforms some host cells into cancerous cells	Histone protein present. Overlapping of genes
Herpes simplex virus (Herpesviridae)	Humans	dsDNA	Mol. wt. 100×10^6 d	Icosahedral	150 nm	Enveloped, complex structure	3 types of mRNA, coding for early and late proteins and DNA
Vaccinia virus (Poxviridae)	Mammals	dsDNA	Mol. wt. 240×10^6 d	Brick-shaped	400 × 240 × 200nm	Complex structure	Source of smallpox vaccine. DNA synthesized in host cytoplasm
HIV (Retroviridae)	Humans (AIDS disease)	ssRNA (2 strands)	Mol. wt. 1.7×10^6 d	Spherical	100 nm	Enveloped, complex structure	Reverse transcription occurs DNA formed gets integrated with host DNA

Plant Viruses:							
Tobacco mosaic virus (Tobamoviridae)	Tobacco	ssRNA "+"	Mol. wt. of virion 40×10^6 d	Helical (rigid rods)	18 × 300 nm	2130 protein subunits Wide host range	Highly stable. Mechanical transmission
Tobacco rattle virus (Tobraviridae)	Tobacco	ssRNA "+"	Virion Mol. wt. 73×10^6 d	Helical (rigid rods)	18 × 188 + 18 × 43 nm	Multicomponent. Both L and S particles needed for infection.	'L' particles code for RNA and 'S' particles for protein Nematode transmitted in soil
Potato virus x (Protexviridae)	Potato	ssRNA "+"	Virion Mol. wt. 35×10^6 d	Helical (flexuous rods)	480–550 × 11–13 nm	Virus-free seed material produced by tissue culture	Enhances symptoms of other viruses in mixed infections Mechanical transmission
Potato yellow dwarf (Rhabdoviridae)	Potato	ssRNA "–"	Mol. wt. 3.5×10^6 d	Helical (bullet shape)	70 × 180 nm	Envelope with spikes present	Leafhopper transmitted
Tobacco ringspot virus (Nepoviridae)	Tobacco	3 kinds of ssRNA	Mol. wt. 1.3, 2.4 2.8×10^6 d	Icosahedral	28 nm	3 types of particles. 2 are needed for infection	RNA sediments at 125, 95 and 50 S. Nematode transmitted in soil
Rice dwarf virus (Reoviridae)	Rice	dsRNA	Mol. wt. 12×10^6 d	Icosahedral	70 nm	Enveloped (thin lipid membrane)	Segmented genome (12 segments). Leafhopper transmitted
Tomato spotted wilt virus	Tomato	ssRNA "+"	Virion Mol. wt. 10×10^6 d	Icosahedral	70–80 nm	Enveloped (with lipids, protein and carbohydrate)	Divided genome (4 segments). Thrips transmitted. Wide host range
Maize streak virus (Geminivirus)	Maize	ssDNA	Mol. wt. 1.0×10^6 d	Paired icosahedron	18 × 30 nm	One DNA molecule inside	Whitefly transmitted
Cauliflower mosaic virus (Caulimoviridae)	Cauliflower	dsDNA	Mol. wt. 4.4×10^6 d	Cauliflower	50 nm	Has more than one coat	Aphid transmitted. One of the few dsDNA plant viruses

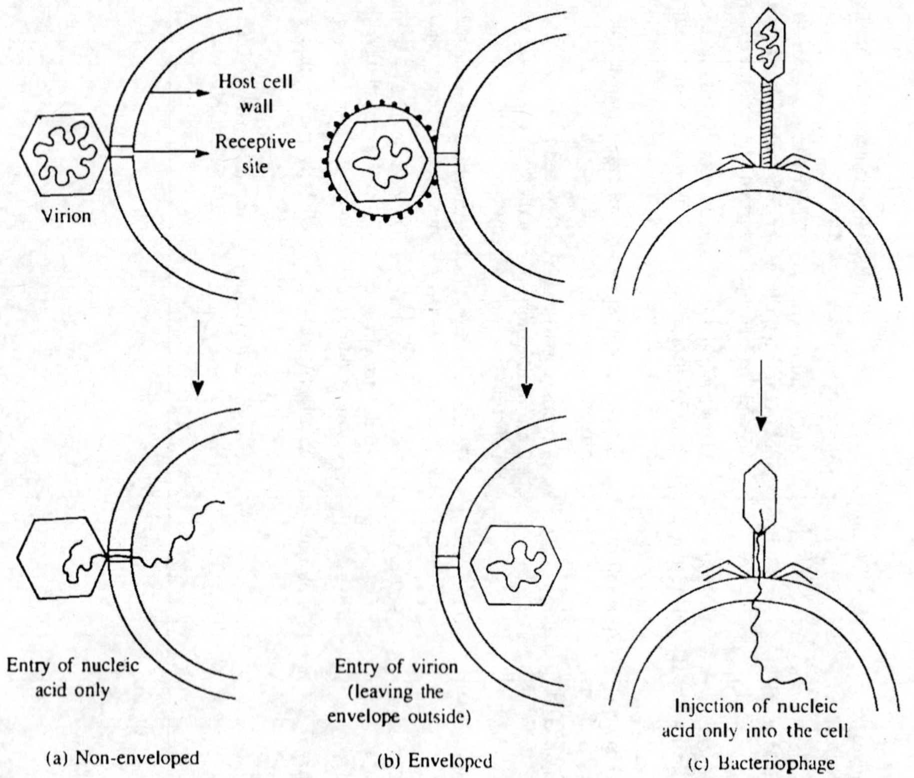

Fig. 16 Types of entry of viruses into host cells.

double-stranded and is known as a **replicative form** (RF). Besides the nucleic acid and coat protein, some viruses also carry accessory proteins and one or more enzymes in the virion.

In some viruses the genome is made up of more than one segment. Four to twelve segments are present and such a genome is called a **segmented genome**. Each segment codes for a *m*RNA. In a few viruses, virions of more than one type which differ from each other either in size or in sedimentation rate occur. Such viruses are called **multicomponent viruses**. Genomes of some of the viruses are so small that all the genes cannot be accommodated in a linear row. Such viruses show **overlapping** of parts of adjacent genes, with different starting points, or coding in opposite directions in the nucleic acid sequences (reading from left to right or right to left). This is a unique feature of some bacteriophages.

SEROLOGY OF VIRUSES

The protein coat or envelope has **antigenic properties** and stimulates the development of **antibodies** when purified or partially purified virus preparation is injected into the blood of an animal (rabbits etc.) The serum, collected from the animal contains the antibodies which precipitate the virus, and is called **antiserum.** When the same or a related virus preparation (*antigen*) is added to the antiserum, precipitation of the antigen and antibodies occurs. The positive reaction indicates the affinity of the antigen and antibody. Thus, when the viruses under test are related, a positive reaction occurs. Hence serological tests have been useful in identifying a virus from a fresh source, or for determining the relationships of viruses from different sources.

REPLICATION OF VIRUSES

The virus modifies the host's genetic mechanism and metabolism, causing it to produce different components of the virus. This results in virus replication. Soon after entering the host cell, the nucleic acid begins to function but the virus cannot be seen until all the components are synthesized. This period is known as an **eclipse period**. After the components are synthesized, the virion is assembled. In a single infection, it occurs only once. Hence, replication is described as a one time process, in contrast to cell division in cellular organisms where the cell number gradually increases by successive divisions. In many bacteriophages, the host cell wall bursts by the activity of **lysozyme** and the progeny virions are released. If the eclipse period is short, fewer virions are assembled and released. If on the other hand, the eclipse period is long, a larger number of virions are released. The number of virions released on bursting of the host cell is known as **burst size**. In a few filamentous bacteriophages, the virion can enter and leave the host cell without bursting of the latter.

The virions of the enveloped viruses acquire the envelope as they are released from the cell. This process is similar to **budding** in bacteria and yeasts. The envelope may be smooth or there may be protein or glycoprotein projections or **spikes** on the surface. They are believed to aid the virion in attachment to receptive sites on the host cell.

A detailed consideration of synthesis of RNA, protein and assembly of the virion of tobacco mosaic virus (TMV) illustrates the process of replication. The TMV coat protein has a single type of protein subunits. They are gradually joined together in a spiral manner. They have points of attachment of RNA towards the inside of the molecule. The nucleotides of the RNA are also assembled in a spiral manner just ahead of the attachment of the protein subunits. Thus RNA is positioned in the hollow of the particle as shown in Fig. 17. The TMV virion is made up of 2130 protein subunits and each subunit is a molecule made up of 158 amino acids. The nucleic acid is single-stranded and has about 6000 bases. The fully assembled virion is 300 nm in length and 18 nm in diameter.

LABORATORY STUDY OF VIRUSES

Viruses, being obligate parasites, cannot be cultured directly on media and indirect methods of cultivation are employed. Plant viruses are maintained

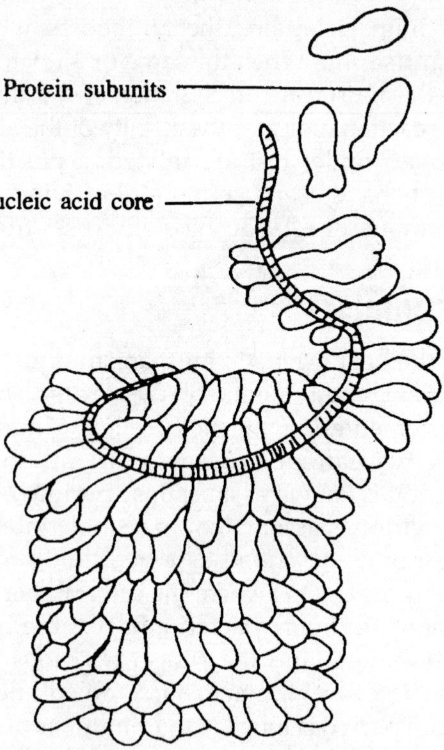

Fig. 17 Assembly of TMV virion.

on susceptible host plants raised in insect-proof wiremesh houses and inoculated with suitable virus preparations. Bacteriophages are maintained on suitable bacterial host cultures. Their physical properties are observed in agar plates seeded with a host culture in which the phage is incorporated. When such a seeded culture develops on agar, clear zones appear in the plate where the bacterial cells are killed. These clear zones are known as **plaques**. The morphology of plaques is useful in identifying the phage. Animal viruses are multiplied either by inoculating healthy host individuals or by inoculating **embryonated chicken eggs**. The virus multiplies in the egg and can be extracted for purification and study. In recent years, animal viruses are being raised in **tissue culture**. Monolayers of cells are developed in a culture and inoculated with the virus. Its multiplication is indicated by the development of plaques, in the same way that phages develop in bacterial cultures.

Purification of virus preparations involves several biophysical and biochemical procedures, including the use of buffers etc., ultracentrifugation and separation techniques. Electron microscope studies have revealed the structure of many viruses. Even the nucleic acid molecules of some viruses are now photographed.

The following account describes the salient features of different types of viruses occurring on plants, animals (including humans) and bacteria, with a brief mention of their importance.

PLANT VIRUSES

Tobacco mosaic virus. The TMV belongs to *Tobamoviridae*. Historically, tobacco mosaic was the first to be recognized as a virus disease. Early research showed that the virus exhibited properties of crystals. Stanley (1935) isolated the virus particles while Bawden and Pirie (1936) showed that the virus is made up of a nucleic acid and a protein. These discoveries have paved the way for rapid and sustained studies on viruses.

The molecular weight of the TMV virion is 40×10^6 daltons. The ssRNA forms about 5% of the virion. The viral RNA itself acts as a messenger for the synthesis of various components. It is synthesized in the host nucleus while the protein subunits are synthesized in the cytoplasm. The virions are assembled in the cytoplasm. The RNA is said to be **polysistronic** in the plant as it codes for all the virion components. TMV has a wide host range and it becomes systemic, producing mosaic symptoms on leaves. It forms **local lesions** in *Nicotiana glutinosa*, a diagnostic criterion for identifying TMV. The TMV develops **high titre** in host tissues, (0.5 to 1.0×10^6 particles per cell) it is highly stable and heat-resistant (thermal inactivation point at 90°C) and spreads through contact. It retains viability even in cured tobacco. There are many strains of TMV and it is believed that they are caused by point mutations in the genome (RNA).

Tobacco rattle virus *(A member of Tobraviridae).* It is a soil-borne virus, transmitted by a nematode, *Trichodorus*, it is **multicomponent**, with long particles of 188 to 197 nm and short particles of 43–114 nm. Both "L" and "S" particles are needed for infection. Its symmetry is helical and its genome is ssRNA. The RNA of "L" particles codes for RNA while that of "S" particles codes for protein. The virions seem to be assembled in mitochondria. The virus produces necrotic symptoms in tobacco and it also infects a few other hosts.

Potato virus X *(Potexviridae).* It occurs in potato, producing mild mottling or no symptoms at all. However, in mixed infections, symptoms of other viruses are enhanced. Because of this danger, tissue culture methods are used to produce PVX-free seed material.

The virion is a long and narrow **flexuous rod**, 480 to 550 × 11 to 13 nm, with helical symmetry. Its molecular weight is 35×10^6 daltons. The genome is ssRNA and it accounts for 6% of the virion.

Tomato bushy stunt virus *(Tombusviridae).* The virion is **icosahedral**, 30 nm in diameter, and has a sedimentation coefficient of 140 S. Its molecular weight is 7.8 to 11×10^6 daltons. There are 185 protein subunits. The genome is ssRNA and it accounts for 17% of the virion. The virus produces mottling and distortion in tomato. Transmission is by contact (mechanical).

Cowpea mosaic virus *(Comoviridae).* The virion is an icosahedron,

30 nm in diameter and has a molecular weight of 7.2×10^6 daltons. It is a **multicomponent** virus with ssRNA, *sedimenting* at 115, 95 and 55 S in three components. The middle and bottom particles are needed for infection and carry RNA coding for different genes. The virus is transmitted by beetles.

Tobacco necrosis virus *(and its satellite virus).* The TNV is **icosahedral** with 28 nm particles. Its molecular weight is 7.5×10^6 daltons and it has ssRNA constituting 17 to 20% of the virion. It causes necrotic mottling and necrosis of veins in tobacco. It is a soil-borne virus transmitted by the zoospores of the fungus, *Olpidium brassicae.*

The TNV satellite virus is a small icosahedral virus (17 nm) which is dependent on TNV for transmission, and is therefore, known as a **satellite virus.**

***Maize streak virus*.** This is a *Geminivirus,* the virion being made up of two modified icosahedrons attached to one another. It is a ssDNA virus, the DNA being circular in configuration. Only one molecule is present in the virion. Each half of the virion has a diameter of 18 × 30 nm.

Cauliflower mosaic virus *(Caulimoviridae CaMV).* Cauliflower mosaic virus is a large icosahedral virus with 50 nm particles and has a dsDNA genome, forming 11% of the virion. It is believed to possess two protein coats. The DNA is a linear double helix but becomes looped on extraction. CaMV is one of the few dsDNA plant viruses. It is transmitted by aphids.

ANIMAL (AND HUMAN) VIRUSES

Polio virus *(Picornaviridae).* This is a small icosahedral virus, with a diameter of 30 nm. (The picornaviruses are so named because of their small size among animal viruses). The genome is a ssRNA "+" type. It acts as *m*RNA and codes for a large protein (polyprotein) which later cleaves into different components. Molecular weight of the protein is 8×10^6 daltons and that of the RNA is 2.6×10^6 daltons.

Poliovirus is water-borne and can survive for long periods in sewage. It cripples the infected individuals by paralysing their limbs. Three antigenic types are recognized and permanent immunity can be achieved by vaccination with oral **sabin vaccine** which is widely used. It is hoped that universal vaccination will completely eliminate the disease by the year 2000 A.D.

European encephalitis virus *(Arbovirus group A).* This is a member of Togaviridae which cause diseases such as yellow fever, dengue fever and Japanese encephalitis. They are mosquito-borne viruses.

The European encephalitis virus is icosahedral, 50 nm in diameter and is surrounded by a **lipid membrane.** The genome is ssRNA "+". It produces a polyprotein (as in picornaviruses). Disease symptoms are headache, fever, joint pains etc. Permanent immunity is acquired after exposure to the virus. The Japanese encephalitis virus produces more severe symptoms which are sometimes lethal. Vaccine is available against this virus.

Rabies virus *(Rhabdoviridae).* Members of Rhabdoviridae infect plants and animals. The rabies virus infects canines and other mammals. Dogs and foxes may show mild or no symptoms. When symptoms develop, the animal becomes aggressive and may attack humans and animals without provocation. This condition is referred to as **furious rabies**. Direct bites or exposure of wounds etc. to secretions spreads the virus. Louis Pasteur developed a vaccine using the attenuated virus. Humans are more susceptible and unless vaccinated in time, the disease manifests as **hydrophobia** resulting in death of the individuals bitten. At present, tissue cultures of human diploid cell lines are used for vaccine production.

The rabies virus is bullet-shaped with an envelope. The virion is 70 × 175 nm in size. The genome is of the ssRNA "–" type with helical symmetry. The virion carries the enzyme **transcriptase**, which helps to transcribe the "+" type mRNA from the viral mRNA, and also a major protein located between the nucleocapsid and the envelope.

Corona virus. This is an icosahedral virus, surrounded by an envelope that has club-shaped projections resembling the solar corona, hence the name. It is one of the causes of the common cold. The genome is the ssRNA "+" type and is helically arranged and segmented. The virion is 100 to 150 nm in diameter.

Influenza virus *(Orthomyxoviridae).* The influenza virus has enveloped and pleomorphic virions, about 80 × 120 nm in diameter with **glycoprotein external spikes.** The spikes are made up of two types of protein, **haemagglutinin** and **neuraminadase**. The virus enters the nasal and pharyngeal region, affecting the mucus membrane by the neuraminadase. Secondary infections by bacteria aggravate the disease, sometimes causing death of the individual. Removal of spikes by proteases renders the virus inactive. The genome is ssRNA "–" which is **segmented**, with 4 to 8 segments. It has 10,000 to 13,000 bases, but different genes occur in different segments, leading to frequent variations. There are three antigenic strains and protection with vaccines is not satisfactory due to variability of the virus.

Mumps and measles viruses *(Paramyxoviridae).* The term **Myxo** refers to mucus, and these viruses affect the mucous membranes (as in Orthomyxoviridae). The measles virus enters the upper respiratory tract and spreads to other endothelial tissues in the tonsils, adenoids, lymph nodes, lungs, liver etc. High fever and skin rashes develop. Individuals once affected, develop permanent immunity.

The mumps virus enters through the upper respiratory tract and enters the pancreas, genital organs and the brain. The usual symptom is swelling of the salivary glands and acquired immunity is permanent.

The paramyxoviruses are enveloped, about 150 nm in diameter and contain a ssRNA "–" type genome. An **endogenous transcriptase** is present. The envelope of the measles virus has haemaglutinin while that of mumps has haemaglutinin and neuraminadase.

Respiratory, enteric orphan virus *(REO Reoviridae)*. This virus is so named because it does not generally cause symptoms. Reoviruses are a large group of dsRNA viruses infecting bacteria, fungi, plants and animals. The reovirus is icosahedral, 75–80 nm in diameter and has two concentric protein shells, with spikes on the outer shell. The genome has 10 segments and each (dsRNA) segment transcribes an ssRNA which acts as mRNA.

Parvovirus *(Parvoviridae)*. This is a small icosahedral virus with ssDNA "–" type genome. The virion is 25 nm in diameter. Members of this group infect mammals, especially rats and dogs. These are the smallest of animal viruses and overlapping of genes occurs.

ARD virus-4 *(Adenoviridae)*. The adenoviruses infect humans, other mammals and birds. The virion is icosahedral, 80 nm in diameter with long, spike-like projections extending from the 12 vertices. The ARD virus 4 causes pharangitis with headache, cough and fever. The genome is dsDNA. The two strands are linear and are said to have terminally redundant sequences.

SV40 virus *(Papovaviridae)*. This simian vacuolating virus (SV 40) has been extensively studied and is used as a vector in genetic engineering. It, and other members of this group (e.g. Polyoma virus) cause tumors in mice. Its genome sometimes gets integrated with the host chromosome transforming normal cells to cancerous cells. The virion is an icosahedron, 45 nm in diameter and possesses a supercoiled dsDNA. **Histone proteins** cover the DNA (as in eukaryotes).

Herpes simplex virus *(Herpesviridae)*. This virus causes fever blisters in humans. It spreads through water or through oral and eye secretions. It infects infants and is often carried without producing symptoms (latent state). Virus multiplication is induced when body defences become weak and symptoms appear. The virion is large and enveloped, about 150 nm in diameter including the 30 nm thick envelope. Spikes are present on the glycoproteinaceous envelope. The capsid is icosahedral and there is an amorphous material between the envelope and the icosahedron. The genome is a large, linear dsDNA molecule.

Vaccinia virus *(Poxviridae)*. The pox viruses are the largest viruses, just visible under a light microscope. Smallpox, caused by the *variolavirus* is now completely eradicated with the use of vaccine, originally developed by Edward Jenner in 1798, much before the nature of the disease was known.

Jenner used cowpox as vaccine but in the recent past, vaccinia virus (*variola minor*) is used as the vaccine. This latter virus is brick-shaped and 400 × 240 × 200 nm in size. There is a lipoprotein envelope with numerous rod-like projections on the outside. The capsid is dumb-bell shaped. Cylindrical pegs are seen between the capsid and the envelope. Two lateral bodies are also seen. The genome is a dsDNA, which transcribes for 30 polypeptides, a transcriptase and some other enzymes.

HIV *(Retroviridae).* The human immunodeficiency virus (HIV) is the causative agent of AIDS (Acquired immunodeficiency syndrome) which has become a worldwide problem in recent years. It attacks the immune system, especially T-cells and macrophages in the blood as a result of which the patient gradually loses resistance and succumbs to other diseases. It is transmitted through direct contact of blood. Hence, homosexuals and drug users are more vulnerable. Transfusion of blood collected from HIV positive people is also an important source.

The retroviruses are also oncogenic (cancer producing), implicated in some cancers. They are all unique in that their genome is RNA but they replicate through a DNA intermediate. This is coded by the enzyme **RNA-dependent DNA polymerase (reverse transcriptase)** and hence the name **retrovirus.** The RNA is made up of two parallel single strands. The virion is enveloped and complex. It is about 100 nm in diameter. The lipoprotein envelope has spikes made up of two proteins. There is an inner protein shell enclosing an icosahedral capsid. The core consists of several enzymes including reverse transcriptase, besides the two RNA strands. The DNA intermediate gets integrated with the host chromosome, hence control of the HIV with vaccines has been difficult. Much current research is engaged on this aspect. One approach is to develop a **ribozyme**, an enzyme that breaks up viral RNA so that the DNA intermediate is not formed. Figure 18 illustrates the structure of HIV and the process of integration with the host chromosome.

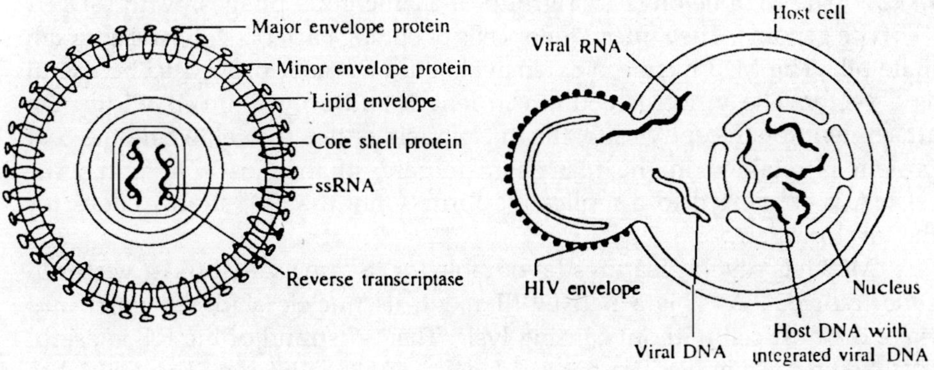

Fig. 18 Structure of HIV and the process of its integration with host chromosome.

Bacteriophages

Bacteriophages were discovered around 1917. With the development of physico-chemical techniques for the study of macromolecules, they came into focus on account of their minute size, replication in bacterial cultures, and their small genomes, making them most suitable for studies on molecular genetics. The genomes of many of them have been mapped. Some of them proved to be useful as vectors for transfer of genes between organisms. The phages infecting *Escherichia coli* are most extensively studied. The naming of phages is by assigning a Greek or English alphabetical letter and a number to them. The following account describes the salient features of some well understood phages.

Figure 19 illustrates genome maps of phages MS2, $\phi \times 174$ and T7.

MS2. The MS2 occurring on *E. coli* belongs to a group of small icosahedral phages with ssRNA genomes. These RNA phages infect only donor cells of enteric bacteria, as their entry is through male pili. The virion of MS2 is 26 nm in diameter and made up of 180 protein subunits. The ssRNA is "+" type, has 3569 nucleotides and shows several loops in configuration. It gets attached to the host ribosomes, acting as mRNA, and codes for four or more proteins (maturation protein, coat protein, lysis protein) including RNA replicase. The gene coding for lysis protein overlaps the genes coding for coat protein and replicase. The replicase is partly coded by the viral RNA, and partly by the host genome.

$\phi \times$ 174. This belongs to a group of phages with icosahedral symmetry and possessing ssDNA genome. The *$\phi \times 174$* occurs on *E. coli* and the virion is 25 nm in diameter, with 60 protein subunits. There are proteinaceous spikes on vertices of the virion. The DNA is circular, with 5386 nucleotides. On entry into a host, the DNA is converted into a double-stranded, supercoiled replicative form, with closed and circular configuration. One strand of the RF acts as mRNA for proteins and the other strand codes for ssDNA (for the viral progeny).

M13. The M13 belongs to a group of filamentous phages, with ssDNA "+" type genome. They infect donor cells of enteric bacteria, entering through male pili. The M13 has been extensively studied, as it proved to be useful as a vector. The virion is 860 nm in length and only 6 nm in width. On attachment to a receptive site, the nucleic acid enters the cell while the coat protein is retained in the plasma membrane of the host. On entry, the ssDNA is converted to a replicative form. Only the "+" strand codes for the viral progeny.

M13 has several features favourable for its use as a **cloning vector** in gene transfers. As it is a narrow filament, its nucleic acid can enter and leave the host cell without causing lysis. The "–" strand of the RF is useful as a cloning vector, as it is not needed for viral replication. The DNA has many restriction sites at which foreign DNA can be inserted.

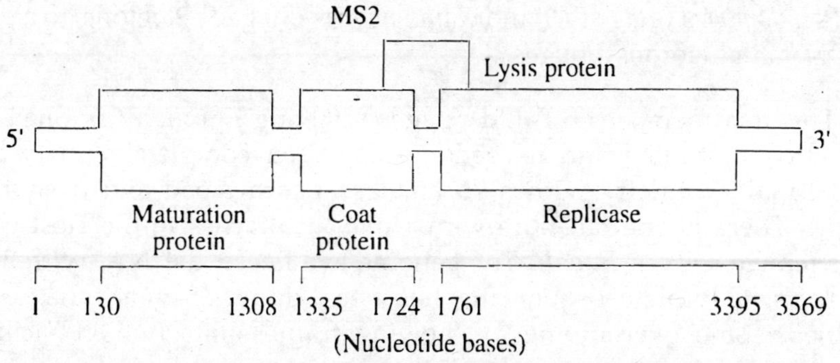

(a) Genome map of MS2 showing overlapping of gene coding for lysis protein

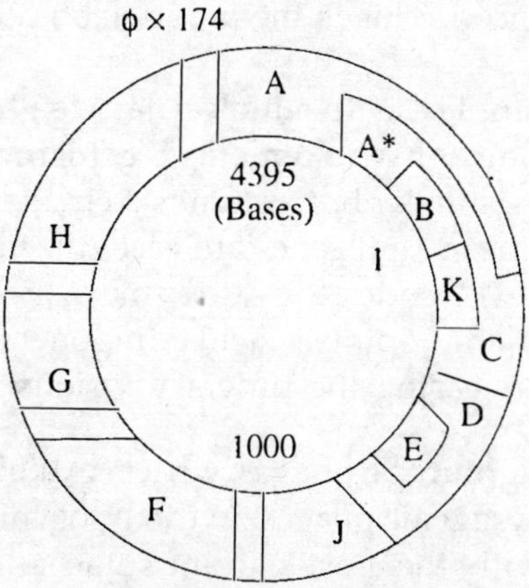

A Synthesis of RF form
A* Shut off of host DNA synthesis
B Formation of capsid precursors
C DNA maturation
D Capsid assembly
E Host cell lysis
F Major capsid protein
G Major spike protein
H Minor spike protein
J Core protein
K Synthesis of a protein

(b) Map of φ × 174 showing overlapping of several genes

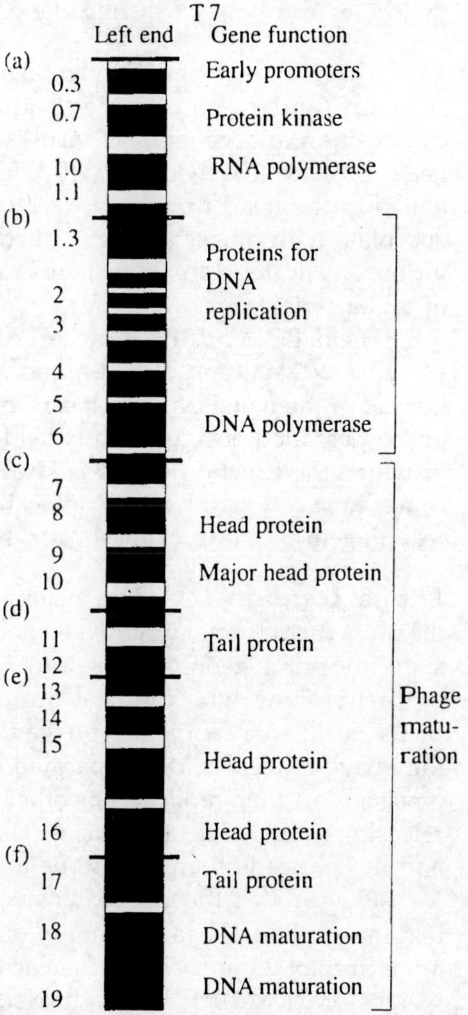

(c) Map of T 7 genome (linear)

Fig. 19 Genome maps of bacteriophages.

Pseudomonas phage *Pf* 1 and *Xanthomonas* phage *Pf* 2 belong to a group of ssDNA filamentous phages.

T7. This is a member of **T-odd** phages infecting *E. coli*. T7 is one of the earliest to be studied and its genome has been completely analysed. It shows **binal symmetry** with a 75 nm icosahedral head and a short tail (20 nm). There is one tail fibre with which it attaches to the host cell. It infects female cells of ***E. coli***. The genome is a linear, dsDNA, with 39,936 base pairs. Complete sequencing has been done. 25 genes have been identified, some overlapping or showing frame shifts (which facilitates coding for long proteins). DNA replication starts at some point on the molecule and proceeds bidirectionally. Hence, bubble-shaped or Y-shaped molecules can be seen during the eclipse period.

T4. This is a member of **T-even** phages of *E. coli*. They have a very complex structure and have provided valuable information on the structure, composition and genetics of viruses. The T4 has an extended icosahedral head (pronate icosahedron), 110 × 85 nm, and a helical tail (110 × 25 nm) joined to the head through a neck (which has a collar and whiskers). An end plate with pins and long, jointed tail fibres forms the lower end of the virion. In all, there are 25 proteins in the virion. This is the most elaborate of all bacteriophages (Fig. 15).

The dsDNA of T4 is quite large and complex, with a molecular weight of 120×10^6 daltons. It is unique in having **5-hydroxymethyl cytosine**, instead of the usual cytosine base. The DNA is linear but assumes a circular and supercoiled position tightly held in the head. 160 genes are recognized, of which the function of 120 is known. The T4 codes for lysozyme which causes host **cell wall lysis** (by attacking its diaminopimelic acid component) resulting in bursting of the host cell thus releasing the progeny virions.

Phage Lambda (λ). This belongs to a group of phages which exhibit the property of **lysogeny**, and are known as lysogenic phages. In the lysogenic state, the viral genome gets integrated with the host chromosome and replicates along with the host during cell division. In this condition, the phage is called a **temperate phage** or **prophage.** Under certain conditions, the phage nucleic acid is separated from the host chromosome, becoming a plasmid. It then replicates as other phages do and causes lysis of the cell, releasing progeny virus particles. This state is known as the **lytic** condition and the phage is then said to be a **virulent phage.**

Phage "λ" is the most extensively studied lysogenic phage. The virion has an icosahedral head, 65 nm in diameter, and a helical tail (150 nm) with a tail fibre of 23 nm. The nucleic acid is a linear dsDNA, with a molecular weight of 31×10^6 daltons. It assumes a circular form when occurring freely in a host cell. In this form it has been completely sequenced. There are 48503 base pairs. Genes responsible for both lysogenic and lytic actions are present in the genomes and which action prevails is dependent on the cell environment.

Phage λ has been a very useful vehicle as a **cloning vector** as it is possible to form **hybrid DNA** by replacing parts of its DNA with foreign DNA. This is possible because the entire central part of its genome is not necessary for its replication but is required only to induce lysogeny (integration with host chromosome). Figure 20 illustrates integration of the λ genome with the host chromosome. It has been possible to introduce up to 35 to 40 eukaryotic genes into its central portion. Such a hybrid is known as a **cosmid**. When such a **DNA-Cos hybrid** is introduced into *E. coli* it behaves as a plasmid. Thus, eukaryotic genes can be introduced into bacteria. This is an important aspect of biotechnology. Human health products such as **insulin** and **interferon** are now produced this way.

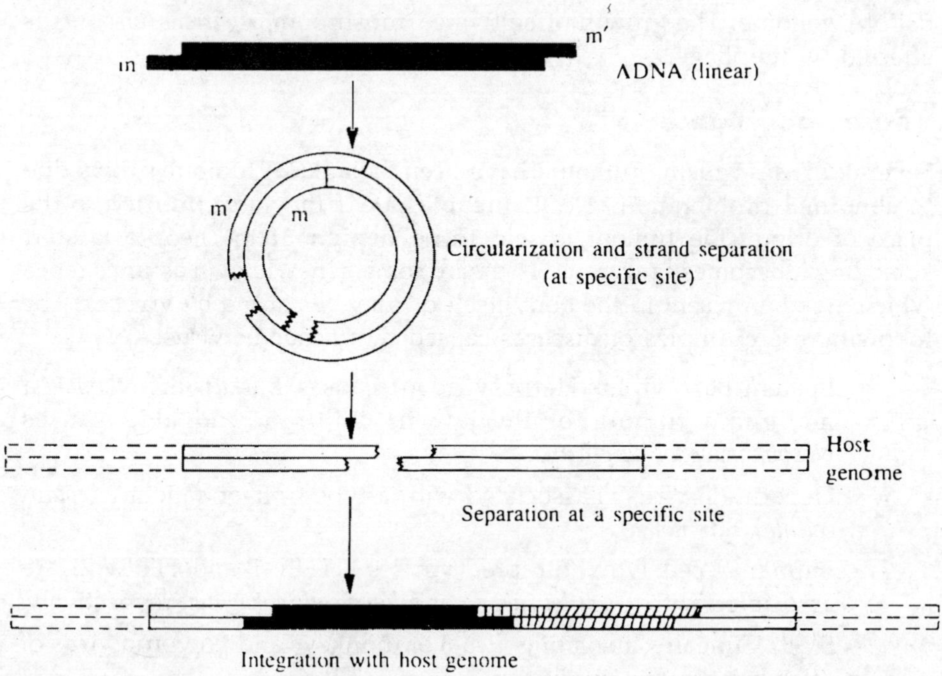

Fig. 20 Integration of phage with host chromosome.

Mu (A transposable phage). This phage is named Mu because it causes mutations in hosts into which it integrates. When it gets inserted into a host chromosome in the middle of a gene, the result is the non-functioning of that gene and the host then behaves as a mutant. During replication, when the phage nucleic acid comes free from the host DNA, it carries a portion of the host DNA with it, causing transposition of genes when it infects a fresh host. Therefore, Mu is called a transposable phage. (Such transfer of genes from one cell to another through a phage is called **transduction.)**

Mu is a large ds DNA phage with **binal symmetry**. The head is about 75 nm and the tail about 110 nm. There are six tail fibres. The genome has about 38 to 39 kilobases. The manner in which the DNA is replicated causes a part of the host genome to get integrated.

The phage Mu also infects *Salmonella* and *Citrobacter*.

Phage ϕ6. This phage is the only dsRNA phage discovered so far. It infects *Pseudomonas phaseolicola*. It is also unusual in that the icosahedral capsid is surrounded by a membranous envelope. With the envelope it is about 80 nm in diameter. There is an RNA polymerase and three segments of dsRNA genome. Each segment codes for an mRNA.

Fungal phages. Phages have been isolated from higher fungi, particularly *Aspergillus, Penicillium* and *Agaricus*. They have isometric particles and dsRNA genome. The growth of cultivated mushroom, *Agaricus bisporus* is affected by its phage.

Viruses and Cancer

Several viruses causing tumours have been isolated. A tumour grows due to abnormal proliferation of cells **(neoplasia).** If they are confined to the place of origin, the tumour is said to be **benign.** If the neoplasia start spreading it becomes **cancerous.** There are some genes, known as **oncogenes** which may be present in the body itself or may be caused by viruses. The following are examples of diseases caused by oncogenic viruses.

- Epstein-Barr virus (Herpesviridae) causes *Burkitts lymphoma*, a malignant tumour of the jaw in children, and also causes *Nasopharyngeal carcinoma.*
- Hepatitis B virus is associated with a type of liver cancer *(Hepatocellular carcinoma).*
- Human T-cell lymphotrophic viruses (HTLV-1 and HTLV-2) are associated with *T-cell leukaemia* and *hairycell leukaemia*, respectively.
- SV40 (Simean vacuolating virus) of monkeys and Polyoma virus of mice are tumour producing.

When and how oncogenes are stimulated to transform normal cells to cancerous cells is not well understood. However, exposure to carcinogenic chemicals, reduction in immune system activity, combination of adverse factors etc. are implicated.

5

NUTRITION AND GROWTH OF BACTERIA

About 80 to 90 per cent of the bacterial cell is made up of water. The solid matter consists of hydrogen and oxygen (largely derived from water) and carbon, nitrogen, phosphorus and sulphur. Besides these, other macronutrients present are potassium, calcium, magnesium and iron. Manganese, molybdenum, cobalt, copper, zinc and nickel though required in minute quantities are important in metabolism. They are known as **micronutrients** or **trace elements**. Some groups of microorganisms require additional elements: for instance diatoms require silica, and microorganisms occurring in saline habitats need sodium. The principal functions of the nutrient elements are listed in Table 7.

SOURCES OF NUTRIENTS

Carbon

All the organic substances contain carbon, hence, organisms are classified into nutritional groups on the basis of the source and method by which their carbon requirements are met.

1. Photoautotrophs use CO_2 as the principal carbon source, employing light energy. Algae and cyanobacteria possess photosystems I and II and oxygen is evolved in photosynthesis (oxygenic photosynthesis). Photosynthetic eubacteria possess only photosystem I and oxygen is not evolved in the process (anoxygenic photosynthesis).

2. Photoorganotrophs use light as energy source but need some organic compound such as acetate as a source of carbon. Certain photosynthetic eubacteria belong to this group.

3. Chemoautotrophs (Lithotrophs) use CO_2 as the source of carbon and obtain energy by the oxidation of reduced inorganic substances like ammonia, reduced forms of sulphur and ferrous iron. Certain eubacteria show this capacity and they can be cultured in strictly mineral media.

4. Heterotrophs comprise the majority and these microorganisms depend on organic carbon sources in nature for satisfying both their energy requirements and synthesis of carbon compounds in the cell. They are further subdivided into two categories. One group occurs on substrates

Table 7 Macronutrients, Micronutrients and their Functions in Metabolism

Element	*Source*	*Function in metabolism*
Macronutrients:		
C	Organic compounds, CO_2	Main constituents of cellular materials
O	O_2, H_2O, CO_2, organic compounds	
H	H_2O, H_2, organic compounds	
N	NH_4, NO_3, N_2, organic compounds	
S	SO_4^{2-}, HS, SO, $S_2O_3^{2-}$	Constituent of cysteine, methionine, thiamine pyrophosphate, coenzyme A, biotin and L-lipoic acid
P	H PO_4^{2-}	Constituent of nucleic acids, phospholipids, and nucleotides
K	K^+	Principal inorganic ion in the cell, cofactor of some enzymes.
Mg	Mg^{2+}	Cofactor of many enzymes, present in cell walls, membranes and phosphate esters
Ca	Ca^{2+}	Cofactor of enzymes, present in exo-enzymes, endospores
Fe	Fe^{2+}, Fe^{3+}	Present in cytochromes, ferredoxins and other non-sulphur proteins, cofactor of some enzymes
Micronutrients		
Zn	Zn^{2+}	Present in RNA and DNA polymerases, alcohol dehydrogenase, alkaline phosphatase and aldolase
Mn	Mn^{2+}	Present in superoxide dismutase, and cofactor of some enzymes
Mo	MoO_4^{2-}	Present in nitrogenase, nitrate reductase, and formate dehydrogenase
Cu	Cu^{2+}	Present in cytochrome oxidase and oxygenases
Co	Co^{2+}	Present in coenzyme, B12-containing enzymes
Se	Se O_3^{2+}	Present in glycine reductase and formate dehydrogenases
Ni	Ni^{2+}	Present in urease and required for autotrophic growth of hydrogen-oxidizing bacteria
Na^+ and Cl^-	NaCl	Required by halophilic bacteria

such as remains of dead organisms, which they decompose to derive their nutrients. They are known as **saprobes** (saprophytes), while those occurring on or inside living organisms and feeding on them are known as **parasites.**

Nitrogen

Most organisms obtain nitrogen in the oxidized form of **nitrate**. Many bacteria can also utilize ammonium salts. Nitrate is reduced to ammonia before its assimilation into cell constituents by the process of **assimilative nitrate reduction.** Nitrogen requirements can also be met from organic nutrients such as amino acids which are in reduced forms. Carbon and sulphur present in these substances also serve as nutrients. Some bacteria can also utilize molecular nitrogen present in the atmosphere, by the process known as **nitrogen fixation.** They possess special enzymes known as **nitrogenases.**

Sulphur

Sulphate is the principal source of sulphur in many organisms. Reduced forms of sulphur present in organic compounds, such as sulphur-containing amino acids, are also utilized by many microorganisms.

Other Elements

Inorganic compounds present in the environment and those released in decomposition of organic substrates are the principal sources of other major nutrient elements and micronutrients. **Phosphorus** that is bound in organic compounds is released as phosphoric acid during decomposition and is taken up by most organisms in that form.

Growth Factors

Apart from the macro- and micro-nutrients, some microorganisms require additional organic compounds, usually in very small quantities, which are essential for their metabolism. These accessory compounds are known as **growth factors**. They include a wide range of compounds such as vitamins, amino acids, purines and pyrimidines, sterols etc. Vitamins and other growth factors and their functions are listed in Table 8.

ENTRY OF NUTRIENTS INTO THE CELL

The cell membrane plays an important role in the transport of solutes into the cell, where they can be maintained at concentrations higher than in the external environment. The entry of solutes may be through **passive diffusion** in response to the difference in concentration across the membrane. Water and certain gases (e.g., N_2 and O_2) move this way. Certain specific membrane proteins mediate the transport of certain compounds which are otherwise impermeable. This process is known as **facilitated diffusion** and those

Table 8 Vitamins and Other Growth Factors and their Functions

Compound	*Functions in metabolism*
Thiamine (B_1)	Thiamin pyrophosphate is the prosthetic group in decarboxylases, transaldolases and transketolases
Riboflavin (B_2)	Precursor of FMN and FAD which are prosthetic groups of flavoproteins
Pyridoxin (B_6)	Pyridoxol phosphate is a coenzyme for transaminases and amino acid decarboxylases
Nicotinic acid	Precursor of NAD and NADP which are coenzymes of dehydrogenases
Cyanocobalamin (B_{12})	Coenzyme involved in rearrangement reactions
Pantothenic acid	Precursor of coenzyme A and prosthetic group of acylcarrier proteins
Folic acid	Tetrahydrofolic acid is a coenzyme involved in transfer of one-carbon unit
Biotin	Prosthetic group of enzymes catalysing carboxylation reactions
P-aminobenzoic acid	Precursor of tetrahydrofolic acid
Lipoic acid	Prosthetic group of pyruvate dehydrogenase complex
Hemin	Precursor of cytochromes
Vitamin K	Precursor of menaquinone, which functions as electron carrier
Coenzyme M	Coenzyme involved in methane formation.

proteins are known as **carrier proteins**. The entry of solutes against concentration gradient occurs by the mechanism known as **active transport**. This is also similar to facilitated diffusion, but involves the expenditure of metabolic energy. Membrane-bound ATPase is known to take part in this process. Certain substances, which can be carried across the membrane, get chemically modified on entering the cell thus accumulating against the concentration gradient. This is known as **group transport**.

A number of macromolecules, which are common substrates for microorganisms (e.g., starch and cellulose) cannot be transported directly across the membrane. They are hydrolysed outside the cell, by **extracellular enzymes** secreted by the organism, prior to transport.

FACTORS AFFECTING MICROBIAL GROWTH

Temperature. The rate of chemical reactions is determined by the temperature and for living organisms, a suitable range of temperatures influences the activity. For a given organism, highest growth rate is maintained at a temperature designated as the **optimum temperature**, provided there are no other limiting conditions. Growth stops at the **maximum temperature** beyond which cellular constituents are damaged. Maximum temperature is usually a few degrees above the optimum. Growth rate gets reduced as the temperature is reduced and stops altogether below a certain point, which is known as the **minimum temperature**.

The optimum, maximum and minimum temperatures vary for different organisms but for most the range is close. The organisms within the 15 to 45°C range are called **mesophiles**, those between 0 to 20°C, **psychrophiles** and those within the 45 to 75°C range are termed **thermophiles**. Among thermophiles, some can grow at mesophilic temperatures and are said to be **thermotolerant**, while true thermophiles require temperatures above 45°C for growth.

Oxygen. Molecular oxygen is required for a majority of organisms. However, there are some groups of bacteria that can grow in the absence of oxygen. These are the **anaerobic bacteria**. Among them, some are capable of growth in either the presence or absence of oxygen. These are called **facultative anaerobes**. On the other hand, some are inhibited and even killed in the presence of molecular oxygen. They are known as the **obligate anaerobes**. It has been found that they lack the enzyme **superoxide dismutase** which is required to convert a superoxide ion formed in metabolism, to oxygen to remove its toxicity. The organisms that require molecular oxygen at or near atmospheric levels (about 20%) are the **aerobes**. Some organisms which occur under conditions of very low oxygen tension (such as root surfaces in soil) are known as **microaerophils**.

pH value. The hydrogen ion concentration in cellular environment influences growth as solubilities of solutes as well as enzyme activities are influenced by pH. Most microorganisms function efficiently in media near neutral pH. Growth in culture media sometimes modifies the pH, thus interfering with continuation of growth. Such an effect is overcome by adding **buffers** to the medium. A buffer usually consists of a weak acid and its salt. Such a combination resists changes in pH by maintaining the equilibrium between the two.

There are some microorganisms which lower the pH drastically. For instance, the sulphur oxidizing bacterium *Thiobacillus thiooxidans*, lowers the pH to 2.0 (by oxidising sulphur to sulphuric acid) and can carry out its activity even under these extremely acidic conditions.

Osmotic concentration. Cells maintain higher osmotic pressure compared to the surrounding medium. Many bacteria grow in very dilute solutions where osmotic pressure is low. There are some bacteria called **osmophiles** which tolerate higher osmotic pressures. Higher concentrations of sugars and salts inhibit the growth of other organisms but osmophiles withstand such conditions. Those growing in high concentrations of salts, such as sea water or brine are known as **halophiles**. There are bacteria which occur in saturated salt solutions. They are called **extreme halophiles**.

It is important to maintain constant ionic strength for physiological activity of the cell and this is maintained by the presence of regulatory substances. The concentration of the potassium ion (K^+) plays an important role in this. Halophiles require the presence of sodium ions in the environment for their growth.

GROWTH OF BACTERIA

Growth may be defined as increase in the mass of an organism. Under conditions of nutrition and environment favourable for growth, organisms maintain a nearly constant composition of various constituents. Thus, growth can be measured either by determining the **biomass** of the organism or by determining one of its principal constituents.

Growth phases. The growth of bacteria in a culture medium in which conditions for balanced growth are favourable, changes in cell mass can be related to three or four phases, i.e., **lag phase, log phase, stationary phase** and **death** or **autolytic phase**. When an inoculum is introduced into a fresh medium, increase in cell mass does not occur for some time. This is due to the need for the inoculum culture to adapt to the uptake of ingredients present in the medium. This is the **lag phase**. Soon, the cell mass starts increasing and doubles after every cell division. The time taken for doubling is known as the **generation time**. It varies from about 20 minutes to a few hours for most bacteria. As the population doubles by the end of each generation time, growth at this phase becomes exponential (2, 4, 8, 16 and so on). This is the **log phase.** After some time, the growth rate falls, mainly due to limitation of one or more constituents in the medium, and/or accumulation of toxic or unwanted chemicals, or changes in pH. For a brief period, increase in cell mass is balanced by cell death. This is the **stationary phase**. Soon, a net decline in cell mass occurs, leading to the **death phase**. The typical growth curve of bacteria in culture is illustrated in Fig. 21.

Measurement of growth. The common parameters of growth that are measured are cell mass or cell number for bacteria, fresh or dry weight of mycelium for fungi, and chlorophyll content for algae. Cell mass of bacteria is determined by separating the cells from the medium by filtration or centrifugation and determining fresh weight, or dry weight after appropriate drying. The more common method is the **turbidity method,** which measures the optical density of the cell mass in a unit quantity of medium at a suitable wavelength (usually at 650 nm) in a colorimeter. The cell number is determined by accurately counting the number of cells after transferring a uniformly suspended medium to a counting chamber. A **Neubar's chamber** or a **haemocytometer** is employed, where a known

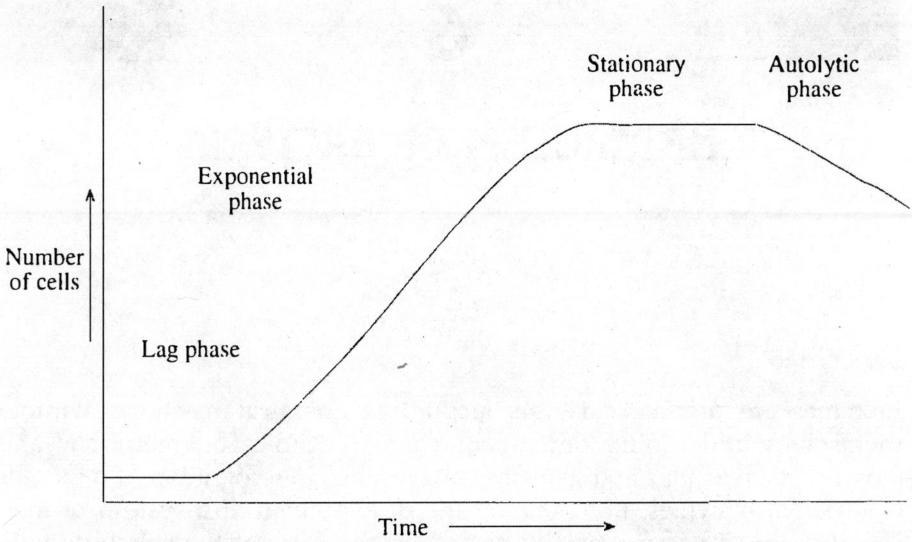

Fig. 21 Typical growth curve of bacteria in culture.

volume of the suspension can be observed under a microscope. Populations of microorganisms from natural substrates can be estimated by diluting the substrate (such as water, milk, soil, sewage etc.) to a required level and plating out on a medium, so that about 50–100 colonies per plate develop (20–30 for fungi). This gives the viable population of any particular group of organisms, depending on the composition of the medium and conditions of incubation.

As mentioned earlier, principal cell constituents are also used to express biomass and DNA, RNA, total protein, peptidoglycan, or chlorophyll content are usually considered for this purpose. The pattern of development of an organism can be better understood if, at the time of harvest, all the cells are at a similar metabolic state. This is achieved through special techniques and the growth under such conditions is called **synchronous growth.**

Populations can also be maintained at a constant level by applying conditions to achieve **continuous culture**. This is of importance in the utilization of microorganisms in industry, in which microbial products are extracted from the culture. The device which permits continuous culture is known as a **chemostat.** In contrast, some products are obtained from **batch cultures**, where the culture is removed at the time of maximum product formation and a fresh batch is started.

6

METABOLISM OF BACTERIA

ENZYMES

Enzymes are organic catalysts facilitating chemical reactions without themselves undergoing permanent change. Studies on metabolic and biosynthetic reactions in organisms have resulted in recognition of hundreds of different enzymes, necessitating the development of a system of their classification. They are usually named on the basis of the substrate with which they react and the type of reaction which they catalyse. They are required in small quantities as they repeat the reaction depending on the conditions.

Enzymes facilitate chemical reactions by lowering **activation energy** of the reaction. It is the energy required (in kcal) to bring all the molecules in a reaction to an active state, so that the chemical bonds can be broken. By lowering the activation energy needed, enzymes increase the rate of reaction. Enzymes are highly specific, catalysing a single reaction or a group of closely related reactions. The enzyme holds the substrate temporarily, through weak bonds, forming an enzyme-substrate complex. The binding is at a small portion of the enzyme, known as the **active site**. This is due to the specific three-dimensional structure of each enzyme, which is a protein, in which the linear chain of amino acids is folded into a secondary and a tertiary structure. Some enzymes have more than one active site.

Small non-protein molecules participate in most enzyme reactions. They are **prosthetic groups**. They are bound very tightly with the enzyme, sometimes permanently, the protein part being known as an **apoenzyme** and together with the prosthetic group as a **holoenzyme**. In some enzymes the activity is associated with a **coenzyme** which is loosely bound to the enzyme. Often a coenzyme may associate with different enzymes. Coenzymes are mostly vitamins.

When physical conditions are favourable, the rate of reaction depends on the concentration of the enzyme and that of the substrate, and the velocity is maximum when the enzyme is saturated with the substrate. The rate of reaction of an enzyme is usually expressed as the **Michaelis-Menten constant,** which is the rate, in moles/litre when the velocity of reaction is at half its maximum. This constant is expressed by the symbol **Km**. More than one enzyme can catalyse some reactions. Such enzymes are known as **isoenzymes**. Km determination is a useful means of identifying them. It is also useful in determining the number of active sites on an enzyme because at the rate of Michaelis constant, half the active sites on the enzyme are filled.

The **metal elements** are also associated with enzymes, as activators. Potassium and iron among the macro elements and zinc, molybdenum, copper, manganese and nickel among micronutrients are associated with various enzymes.

Enzyme activity is regulated in most cases by a **feedback mechanism**. If an end product is synthesized in excess of immediate requirement, it inhibits the first intermediate in the pathway. Other enzymes in the pathway stop functioning for want of substrates. The pathway of isoleucine synthesis illustrates this feedback mechanism.

$$\text{Theonine} \longrightarrow \alpha\text{-oxobutyrate} \longrightarrow \text{Isoleucine}$$

Another mechanism, known as the **cumulative mechanism** operates in some pathways. Here, one or more of the intermediates cause different degrees of inhibition.

The mechanism of attachment of substrate to enzyme at an active site is described as a **lock and key mechanism**. However, it has been shown that many enzymes get complementary shape through interaction with the substrate. Such enzymes show **allosteric property**. Figure 22 illustrates the models showing relation between enzyme and substrate.

In the absence of a substrate, active sites of an enzyme have low catalytic activity but when one of the sites is occupied by a substrate, it brings out conformational changes in the enzyme so that catalytic activity at other active sites also increases. If a negative factor occupies one site, all the sites show reduced activity. Such allosteric control regulates the products in metabolism. For instance, pyruvate dehydrogenase activity is inhibited by NADH or acetyl-CoA, but activated by PEP, AMP and GDP in *E. coli*. PEP carboxylase is inhibited by malate and aspartate (indicating an excess of C_4—dicarboxylic acids) and stimulated by acetyl—CoA (indicating the need for C_4—dicarboxylic acids).

Enzymes are classified on the basis of the substrates whose reactions they catalyse and the type of change that is brought about. Table 9 lists important enzyme groups and the nature of their activity.

OXIDATION-REDUCTION

Oxidation involves the removal of an electron from a substance and reduction involves the addition of an electron. The hydrogen atom consists of an electron and a proton and in many instances both protons and electrons are transferred. When reduction occurs, simultaneous oxidation must also occur since electrons do not have free existence and must be transferred to another substance which gets reduced. The substance releasing the electron is an **electron donor** and the substance receiving it is an **electron acceptor**. Thus, each oxidation-reduction consists of two half reactions.

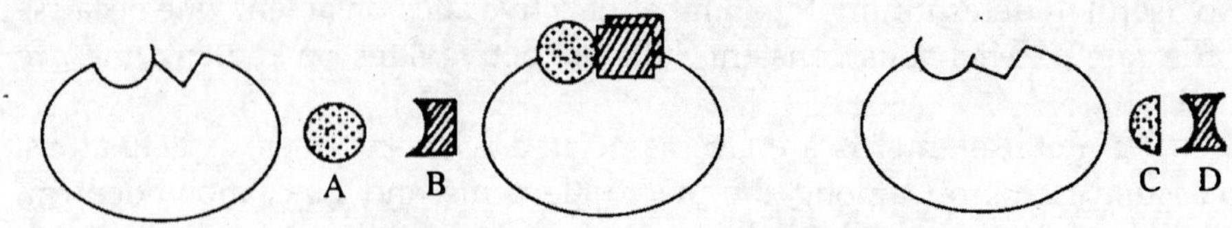

(a) Oxidation of a substrate by enzyme action

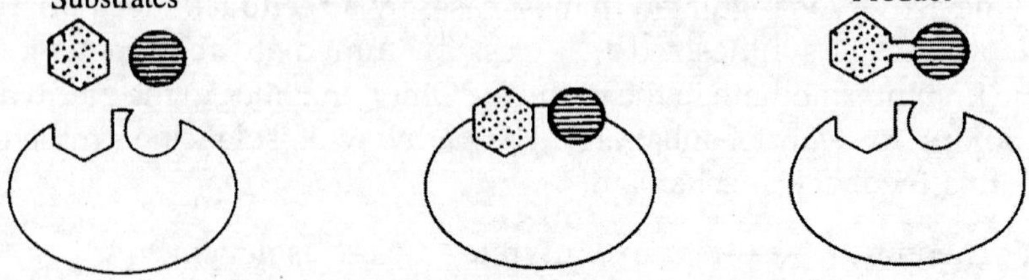

(b) Formation of a product by enzyme action

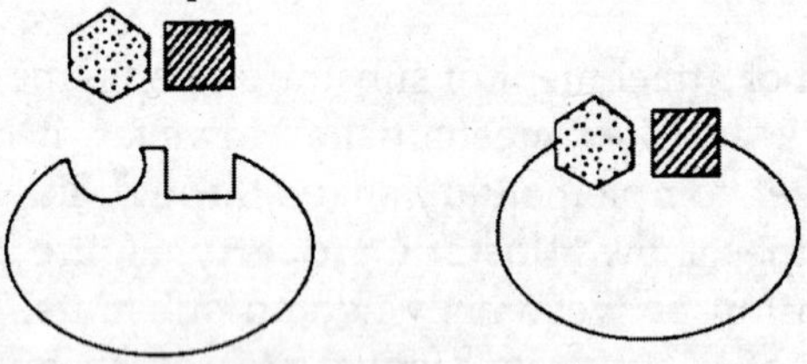

(c) Allosteric effect in enzyme action

A–NAD
B–Substrate (electron donor)
C–$NADH_2$
D–Substrate (oxidized)
E–Allosteric effector
F–Substrate

Fig. 22 Models of enzyme substrate complex.

The passage of electrons from one substance to another is possible because different substances receive electrons at different electrical potentials known as **redox potentials**.

Most molecules either accept or release electrons depending on reduction potential of the substance with which they are reacting. The molecules with more negative redox potential donate electrons to those with more positive potential. The redox potentials are expressed as electron volts, taking hydrogen as standard. Hydrogen has a redox potential of – 0.42 while oxygen has +0.82. Hence the former acts as an electron donor while the latter acts as an electron acceptor. H_2 is oxidized to $2H^+$ while $\frac{1}{2}$ O_2 is reduced to form H_2O.

The transfer of electrons in oxidation-reduction reactions involves some **carrier substances.** Thus, electrons are envisaged as being carried from a **primary electron donor** to a **terminal electron acceptor**. The energy change in the completed reaction depends on the difference in redox potentials of the primary donor and terminal acceptor.

Table 9 Classification of Enzyme Groups

Enzyme group	*Activity*	*Examples*
Oxidoreductases	Oxidation reduction reactions	Glyceraldehyde-3-(p) dehydrogenase, Nitrate reductase (the term oxidase is used only when oxygen is the terminal electron acceptor, e.g. cytochrome oxidase)
Transferases	Group transfer reactions (methyl, glyoxyl etc.)	PEP glucose phosphotransferase 3-Phosphoglycerate kinase, Phosphoglycerate mutase, Dihydrolipoate transacetalase
Hydrolases	Hydrolysis of compounds (cleaving of C–O, C–N, C–C bonds etc.)	β-Galactosidase, α-amylase (endoglucanase)
Lyases	Non-hydrolytic renewal of groups from substrates (cleaving C–N, C–C, C–O)	Fumarase, Fructose biphosphate aldolase, Pyruvate decarboxylase
Isomerases	Isomerization of compounds (geometric changes within a molecule)	Triose phosphate isomerase, UDP glucose epimerase
Ligases (Synthetases)	Joining of molecules	DNA ligase, Acetyl CoA synthetase, Pyruvate carboxylase

ELECTRON CARRIERS

These are membrane-bound and perform the important function of formation of ATP during electron transport from a donor to an acceptor. They include flavoproteins and cytochromes as well as quinones and iron-sulphur proteins. The flavoproteins are proteins containing a derivative of riboflavin (Vitamin B_2). Flavine mononucleotide (FMN), and flavin adenine dinucleotide (FAD) are examples. Cytochromes are proteins with an iron-containing porphyrin ring. The porphyrin ring is known as **heme**. It is made up of four pyrrole rings with the iron atom at the centre. The iron atom can either accept an electron or donate an electron.

$$\text{Cytochrome—Fe}^{2+} \longrightarrow \text{Cytochrome—Fe}^{3+} + e^-$$

There are different cytochromes designated as a, b, c, d, o, etc. They transfer electrons from the one with more negative potential to the one with more positive potential.

Quinones are lipid-soluble substances and can act as electron donors or as hydrogen acceptors. Coenzyme Q (ubiquinone) and menaquinone are examples.

Nicotinamide adenine dinucleotide (NAD^+) and NAD-phosphate ($NADP^+$) are freely diffusible carriers and transfer two hydrogen atoms to the next carrier. NAD^+ is generally involved in **catabolic reactions** generating energy while $NADP^+$ is primarily involved in **anabolic reactions (biosynthetic functions)**.

Ferredoxin and flavodoxin are iron-sulphur proteins with redox potentials close to that of hydrogen, and are hence good electron carriers.

PHOTOSYNTHETIC PIGMENTS AND LEGHAEMOGLOBIN

Chlorophyll-a and bacteriochlorophyll. All chlorophylls have porphyrin ring structure just like cytochromes but the metal that is attached to the four pyrrole rings is **magnesium** instead of iron. Chlorophyll-*a* is present in *cyanobacteria* and *prochlorophytes*. It contains specific side units and a long hydrophobic alcohol side chain, which enables the pigment to adhere to photosynthetic membranes. It absorbs light in the red region, with the maximum at 680 nm. Bacteriochlorophyll-*a*, present in purple and green phototrophic bacteria, has specific side units which are different from those of chlorophyll-*a*. The structure of these pigments as well as that of a cytochrome is illustrated in Fig. 23. Peak absorption of the former is at 850 nm. Slightly different pigments with different absorption peaks also occur. Five different bacterial chlorophylls are known in phototrophic bacteria and a similar number of chlorophylls are present in eukaryotic algae.

In prokaryotes, chlorophyll pigments are attached to thylakoid membranes, in groups of 200–300 molecules. Some of the molecules directly

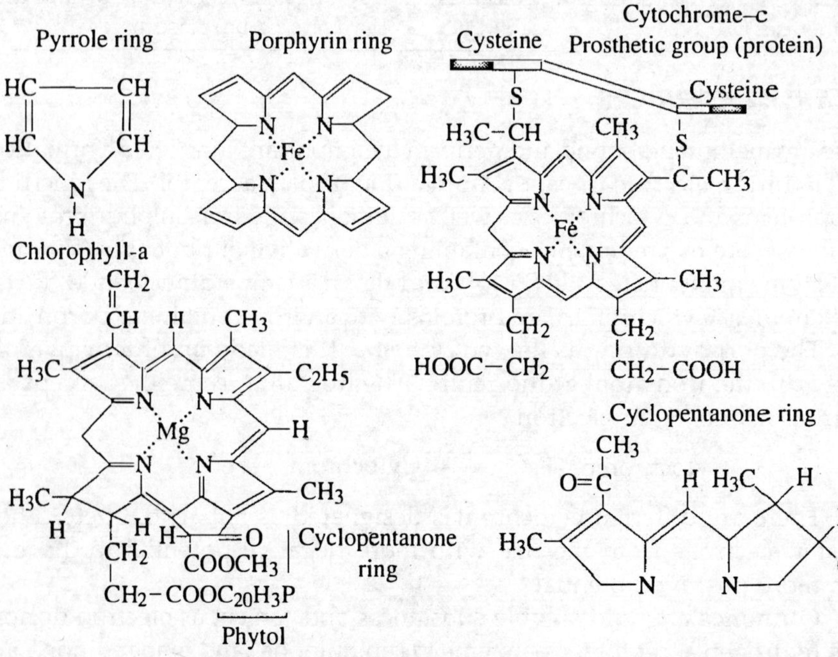

Fig. 23 Structure of cytochromes and chlorophylls.

participate in ATP formation and are known as **reaction centre molecules** while the others are called **antenna** (light harvesting) **molecules.** They transfer photons to the reaction centre.

Carotenoids. Carotenoids are long chain hydrocarbon molecules with conjugated double bonds (double bond alternating with single bond). They are fat-soluble and are embedded in the membrane. They **fluoresce** when excited by light and some of the energy is transferred to chlorophyll. They are usually yellow or red in colour and absorb light in the blue end of spectrum (400–500 nm).

Phycobiliproteins. Phycobiliproteins are accessory pigments found in cyanobacteria (and also red algae). *Phycoerythrin* is a red pigment with absorption peak at 550 nm, and *phycocyanin* is a blue pigment, with absorption at 620–640 nm. They contain open chain **tetrapyrroles bound to protein**. They are associated with thylakoid membranes and transfer light energy very efficiently to chlorophyll and their presence is responsible for the bluish or reddish colour of cyanobacteria. Cyanobacteria are able to grow in very low light intensities also due to their efficient light-harvesting system.

Leghaemoglobin. This is a haemoglobin-like compound occurring in nodules of leguminous plants. It controls oxygen content during nitrogen fixation. It appears that genetic control of leghaemoglobin resides both in the host plant and in the bacterium, the **heme** part being coded by the bacterium and protein part by the host. Oxygen levels are kept low by the heme part with the iron in the centre occurring in oxidized (Fe^{3+}) or reduced (Fe^{2+}) states. The pink or red colour of nodules is due to the presence of this compound.

METABOLISM IN BACTERIA

The biosynthesis scheme starting from glucose has been thoroughly studied in Gram-negative rod shaped bacterium *Escherichia coli*, a facultative anaerobe which can utilize a wide variety of substrates. Under aerobic conditions a part of the substrate is oxidized to CO_2 with oxygen as the terminal electron acceptor. This **catabolic process**, traditionally called respiration, is **exergonic** and results in the formation of ATP, which is required for the biosynthesis of cellular materials. The remaining part of the substrate is utilized for biogenesis of cellular constituents. The latter is an **anabolic process**.

Respiration

Glucose is the commonly employed substrate in the catabolic process. It does not directly enter the cells but is phosphorylated and transported at specific sites in the plasma membrane. Figure 24 illustrates the mechanism of transport of glucose across the membrane. The catabolic process consists

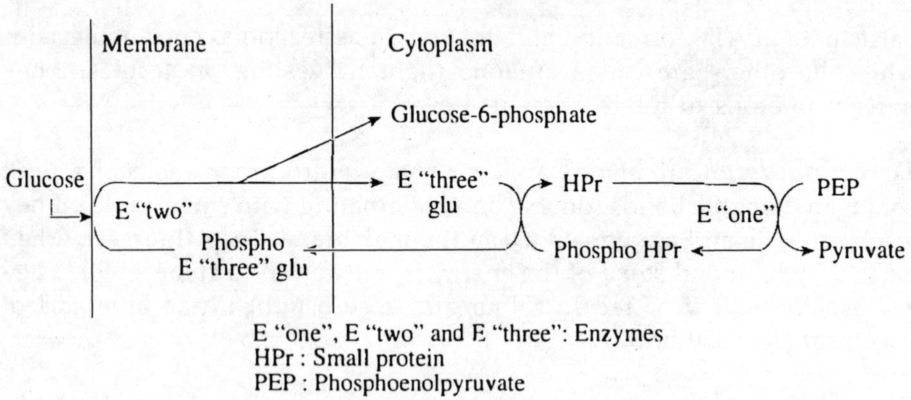

Fig. 24 Transport of glucose into cell.

of three main steps, viz., **glycolysis** (breakdown of glucose to form pyruvate); **TCA cycle** (to produce reduced compounds); and **electron transport** (to carry electrons to oxygen). The three steps together constitute **respiration**.

The Embden, Meyerhof, Parnas pathway (EMP pathway). This is the most commonly occurring pathway for the conversion of glucose-6-phosphate into pyruvate in animals, plants and most bacteria. Figure 25 illustrates the EMP pathway. Fructose-1,6-biphosphate is the characteristic intermediate in this pathway. Fructose biphosphate aldolase splits it into C-3 compounds (dihydroxyacetone phosphate and glyceraldelyde-3-phosphate). The two are in equilibrium due to the action of the enzyme triosephosphate isomerase.

The oxidation of glyceraldehyde phosphate to pyruvate is initiated by two enzymes. The first of these is glyceraldehyde-3-phosphate dehydrogenase. It contains bound NAD which is reduced. The product, 1-3-biphosphoglycerate is hydrolysed by the second enzyme, 3-phosphoglycerate kinase, to 3-phosphoglycerate, transferring its free energy to form ATP from ADP. Thus two molecules of ATP are formed from the two C-3 compounds.

In the next two steps, 3-phosphoglycerate is converted to phosphoenolpyruvate by the action of two enzymes. Phosphoglycerate mutase transfers phosphoryl group from the 3 position to the 2 position, and enolase (phosphopyruvate hydratase) removes water to yield phosphoenolpyruvate. This compound also has a phosphoryl bond with high energy. Conversion of phosphoenolpyruvate to pyruvate yields ATP from ADP by the transfer of this energy. Out of the two molecules of phosphoenolpyruvate, one contributes the phosphoryl group to glucose to form glucose-6-phosphate. Thus, 3 molecules of ATP are generated in this system at **substrate level phosphorylation.** Pyruvatekinase is the enzyme converting phosphoenolpyruvate to pyruvate. Formation of fructose-1-6-biphosphate from fructose-6-phosphate consumes one ATP molecule. Thus **the net gain in EMP pathway is two molecules of ATP per one glucose molecule.**

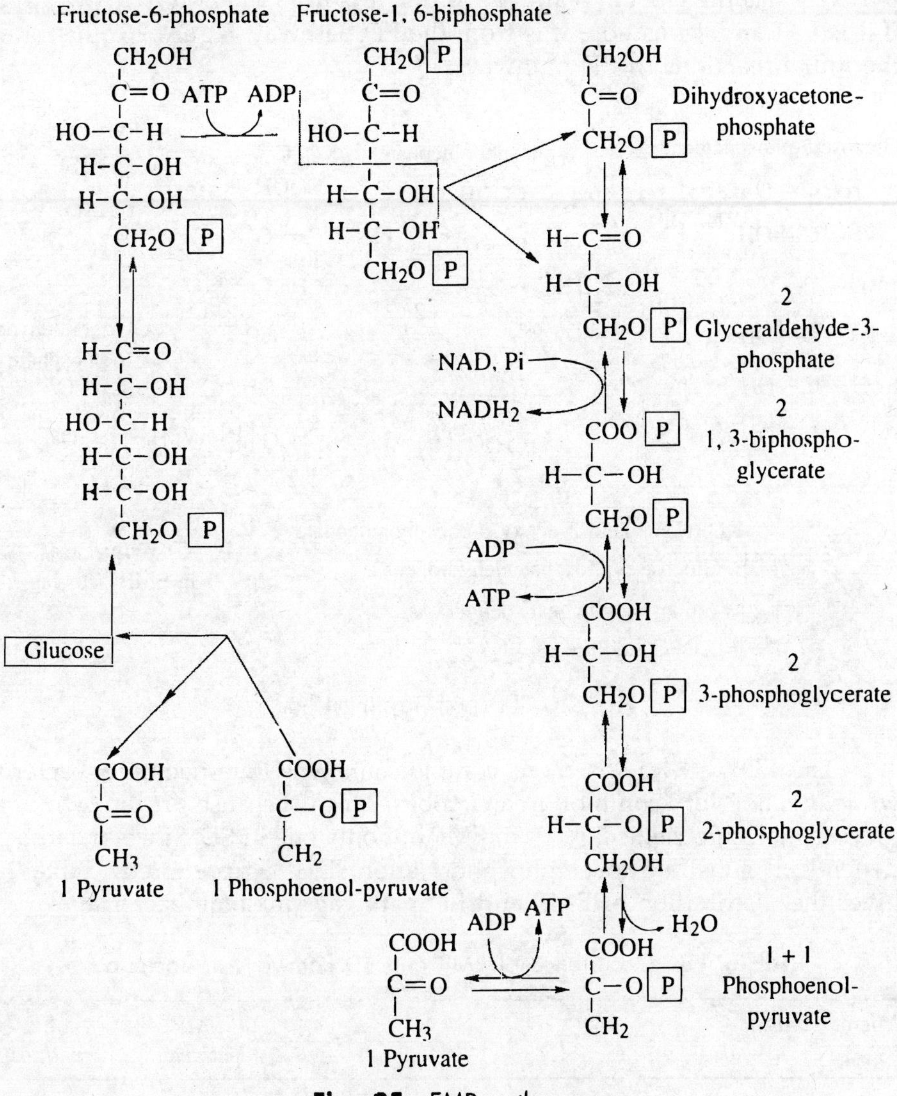

Fig. 25 EMP pathway.

The Entner-Doudoroff pathway (E-D pathway). This second pathway is also widely distributed in microorganisms for carbohydrate breakdown. It was discovered by Entner and Doudoroff in *Pseudomonas saccharophila*. Glucose-6-phosphate is dehydrogenated to yield 6-phosphogluconate, which is converted to one molecule of 3-phosphoglyceraldehyde and one molecule of pyruvate. The enzymes involved are a dehydratase and an aldolase. The 3-phosphoglyceraldehyde is then oxidized to pyruvate as in the EMP pathway. The key enzymes of this pathway are 6-phosphogluconate dehydratase and 2-keto-3-deoxy-6-phosphogluconate aldolase (KDPG aldolase). If high levels of these two enzymes are present, it is likely that the ED pathway is operating. Radioisotope method employing ^{14}C-glucose can also be used. If the resulting pyruvate has the label in 3 and 4 position it is from the EMP pathway and if it is in 1 and 4 position, it is from the ED pathway. Figure 26 illustrates the initial reactions in ED pathway.

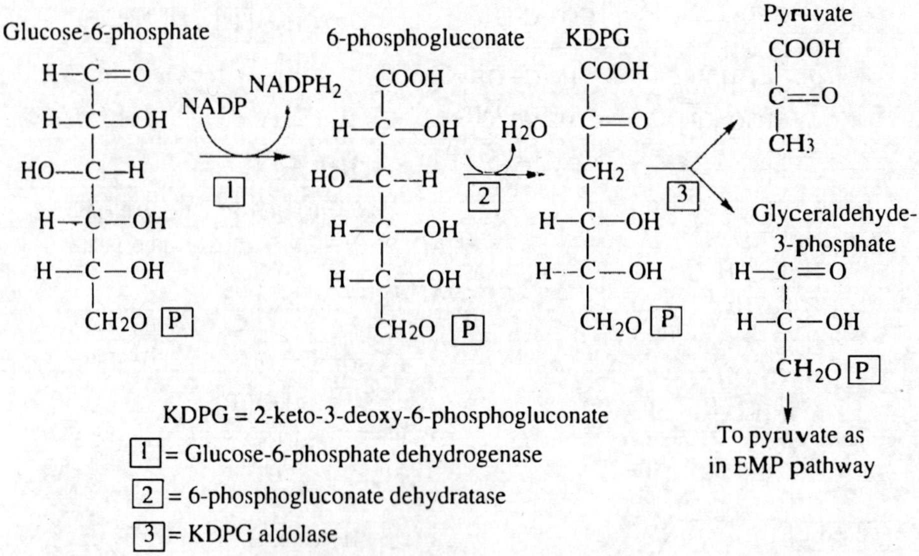

Fig. 26 ED pathway (First half).

The ED pathway is more common among Gram-negative bacteria while it is not very common in anaerobes. This is because breakdown of a glucose molecule yields 2 ATPs in EMP but only one in ED. Since anaerobes depend on substrate level phosphorylation, EMP is preferred. Table 10 gives the distribution of EMP and ED pathways in some bacteria.

Table 10 Occurrence of EMP and ED Pathways in Bacteria

Microorganism	*EMP pathway*	*ED pathway*
Arthrobacter spp.	+	–
Azotobacter chroococcum	+	–
Alcaligenes eutropus	–	+
Bacillus subtilis, B. cereus and other species	+	–
Escherichia coli and other entericbacteria	+	–
Pseudomonas saccharophila, P. fluorescence and other species	–	+
Rhizobium japonicum and other species	–	+
Thiobacillus intermedius and T. ferrooxidans	–	+
Xanthomonas phaseoli and other species	–	+

Besides glucose, other sugars also play important role in bacterial metabolism. *Fructose* is an excellent source for many bacteria. In *E. coli* it is phosphorylated in the same way as glucose. *Lactose* is another widely utilized sugar. Several enzymes are required for metabolizing lactose. They are synthesized in the presence of lactose in the medium. Thus there would be a lag in the utilization of lactose by cells grown previously on glucose. The elucidation of this system was the basis for the discovery of **lactose operon model** by Jacob and Monod. Lactose is converted to **glucose** and **galactose.** The latter is further converted to **glucose-1-phosphate**. The mechanism of lactose uptake is given in Fig. 27.

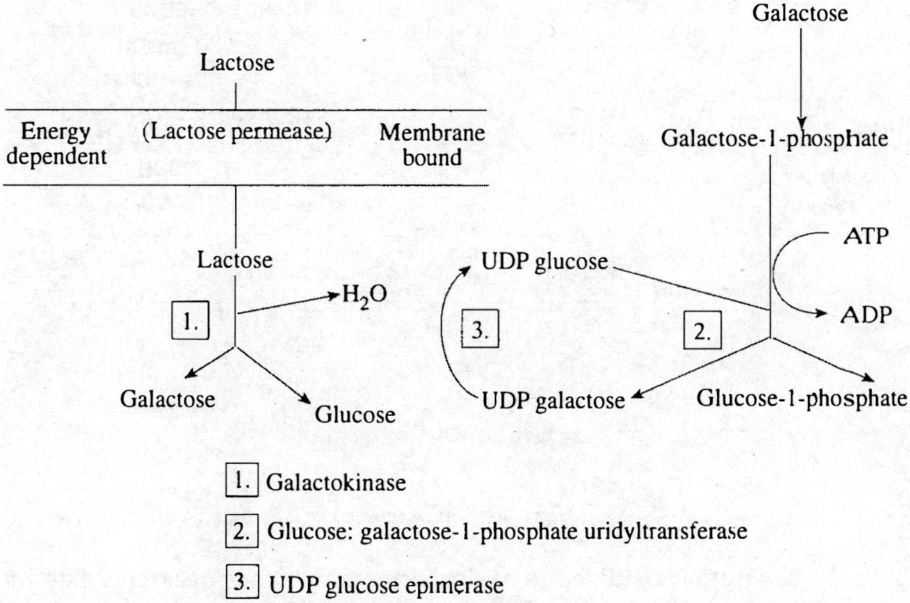

Fig. 27 Uptake of Lactose.

Acetate, malate and pyruvate are also utilized. They are converted to acetyl-CoA and enter the Tricarboxylic acid cycle. Oxidative decarboxylation of pyruvate to acetyl-CoA takes place. The enzyme system is referred to as the pyruvate dehydrogenase system. It consists of 24 molecules each of pyruvate dehydrogenase (E_1), dihydrolipoic transacetylase (E_2) and 12 molecules of dihydrolipoic dehydrogenase (E_3). The E_1 contains thiamine pyrophosphate (TPP). The overall reaction is as follows:

$$CH_3CO\text{—}COOH + CoA + NAD \longrightarrow CH_3CO\text{—}CoA + CO_2 + NADH_2$$

Tricarboxylic acid cycle. The cycle carries out the oxidation of acetyl moiety of the acetyl-CoA to CO_2 with transfer of the reducing equivalent to NAD, NADP and FAD. Oxaloacetate and acetyl-CoA are condensed by the enzyme citrate synthetase to give **citrate**. The subsequent reactions result in the regeneration of oxaloacetate as detailed below, and illustrated in Fig. 28.

1. The citrate is dehydrated to cis-aconitate (which is enzyme bound) which is again hydrated to give **isocitrate**. The enzyme involved is cis-aconitate hydratase.

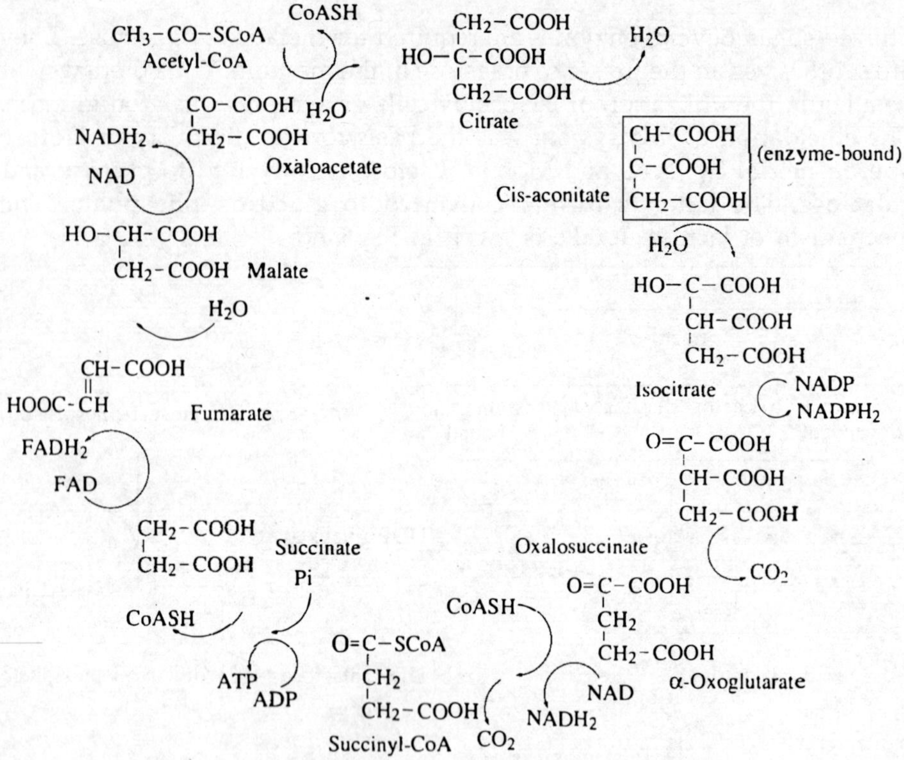

Fig. 28 TCA cycle.

2. Isocitrate is oxidized to oxalosuccinate, which is decarboxylated to α-oxoglutarate, both reactions being carried out by *isocitrate dehydrogenase.*

3. α-oxoglutarate is oxidized to succinyl-CoA (by oxoglutarate dehydrogenase system).

4. The thioester bond of succinyl-CoA is used to generate one ATP from ADP and Pi (substrate level phosphorylation), and formation of **succinate.** The enzyme is *succinate thiokinase.*

5. **Succinate** is oxidized to **fumarate** by *succinate dehydrogenase*, which transfers electrons to FAD, forming $FADH_2$.

6. Fumarate is hydrated to **malate** by fumarase.

7. Malate is oxidized to **oxaloacetate**, by the enzyme L-malate dehydrogenase, with the transfer of electrons to NAD, forming $NADH_2$.

Thus, the overall reaction starting from glucose is

$$C_6H_{12}O_6 + 6H_2O \longrightarrow 24(H) + 6CO_2$$

$$(\text{glucose} + 8NAD + 2NADP + 2FAD + Pi \longrightarrow 8NADH_2 + 2NADPH_2 + 2FADH_2 + 4ATP + 6CO_2)$$

The 24 available hydrogen atoms (theoretically) are transferred to oxygen, in aerobic metabolism, through the electron transport chain, thus releasing NAD, NADP and FAD, to continue the TCA cycle.

Electron transport. The oxidation-reduction potentials at pH 7.0 (OR values, E'o) are – 0.32 and + 0.81, for NAD and oxygen respectively, and the difference between them is the **potential respiratory chain**.

The major components of the respiratory chain are proteins with Oxidation-Reduction potentials lying between those of NAD and oxygen. In the cytoplasmic membranes of bacteria (and mitochondrial membranes of eukaryotes) these proteins are arranged in such a way that the reducing power of $NADH_2$ can flow to oxygen through carriers of increasing OR potentials.

In *E. coli* growing under aerobic conditions, reducing power flows via **coenzyme Q, cytochrome B_{556},** and **cytochrome o, to oxygen.** When oxygen is limited, the cells employ coenzyme Q or **menaquinone** and **cytochrome B_{558}** and **cytochrome d** as carriers. Thus, the respiratory chain is known to be branched in *E. coli* and differs from that occurring in mitochondria where the chain is via coenzyme Q, cytochromes b, e, c_1, c, a and a_3. Figure 29 illustrates the differences in electron transport in mitochondria and *E. coli.*

Oxidative phosphorylation (Electron transport phosphorylation). Part of the energy released during electron transport is utilized for ATP formation (**negotiable energy**). This process is called **oxidative** or **electron transport phosphorylation**.

The exact mechanism of ATP formation is not yet fully understood. Three hypotheses were proposed, i.e., **chemical coupling hypothesis,**

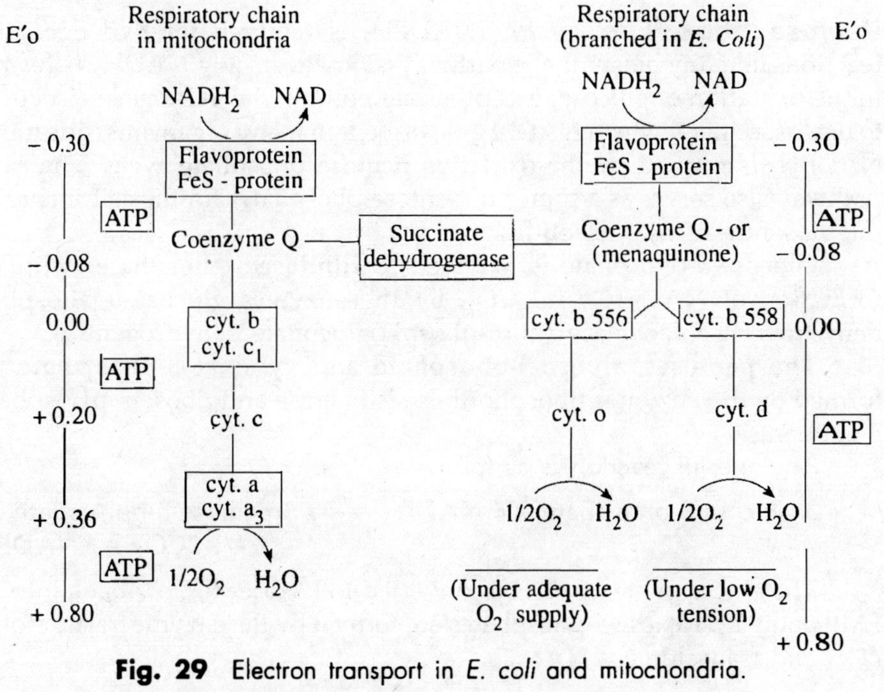

Fig. 29 Electron transport in *E. coli* and mitochondria.

conformational coupling hypothesis and **chemiosmotic hypothesis.** The last one is currently widely accepted. ATP is generated by transporting electrons through a chain of carrier molecules with fixed orientations on the cell membrane, in such a way that a pH gradient and membrane potential are formed by trapping and excretion of protons. This **proton motive force** is taken advantage of by the ATPase (ATP synthetase), allowing the synthesis of ATP from ADP and Pi. Figure 30 illustrates the structure of the ATP molecule and chemiosmotic model of electron transport in oxidative phosphorylation.

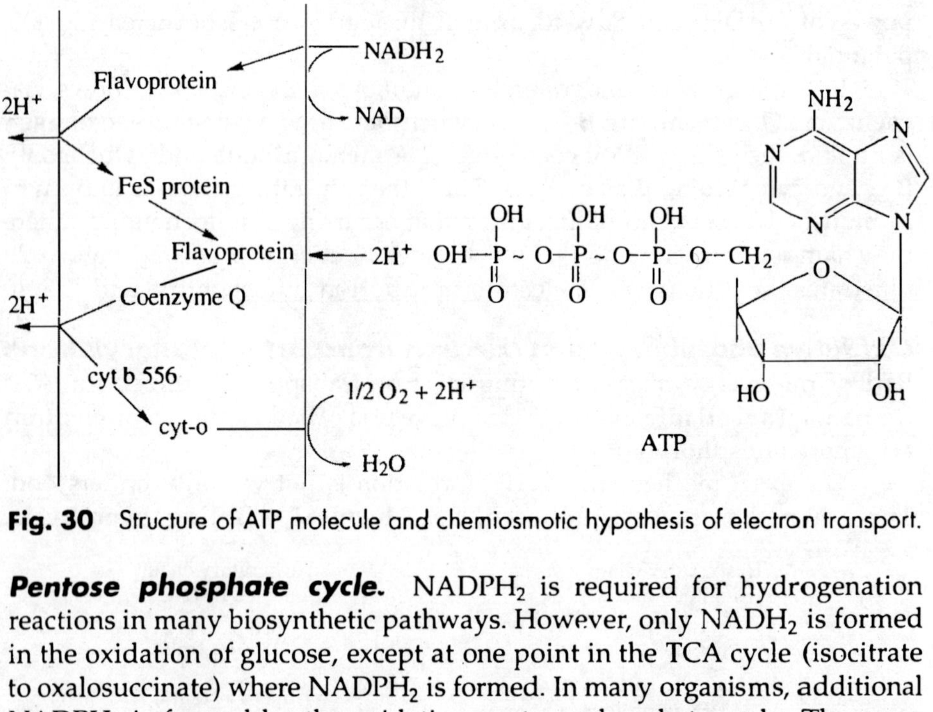

Fig. 30 Structure of ATP molecule and chemiosmotic hypothesis of electron transport.

Pentose phosphate cycle. $NADPH_2$ is required for hydrogenation reactions in many biosynthetic pathways. However, only $NADH_2$ is formed in the oxidation of glucose, except at one point in the TCA cycle (isocitrate to oxalosuccinate) where $NADPH_2$ is formed. In many organisms, additional $NADPH_2$ is formed by the oxidative pentose phosphate cycle. The same pathway also serves as a source of pentose phosphates for the cell including the biosynthesis of nucleotides.

Glucose-6-phosphate is oxidized to **ribulose-5-phosphate,** with the formation of $NADPH_2$ and CO_2, by the enzymes glucose-6-phosphate dehydrogenase, lactonase and 6-phosphogluconate dehydrogenase.

The **pentoses, ribose-5-phosphate** and **xylulose-5-phosphate** are formed by the enzymes phosphoribose isomerase and ribose-5-phosphate-3-epimerase.

The overall reaction is as follows:

$$3 \text{ glucose-6-phosphate} + 6 \text{ NADP} \longrightarrow 3 \text{ pentose-5-phosphate} + 3CO_2 + 6NADPH_2$$

The excess pentoses can be converted to hexoses to flow back into the EMP pathway. Hexose-6-phosphates are formed by the enzyme transketolase (TK) and transaldolase (TA).

The pentose phosphate pathway also produces erythrose-4-phosphate, which is utilized in biosynthesis of aromatic amino acids. In a majority of aerobic organisms, about 20% of glucose flows into the pentose phosphate cycle for fulfilling the requirements of cells for pentose phosphates. Figure 31 illustrates the pentose phosphate pathway (starting from glucose-6-phosphate).

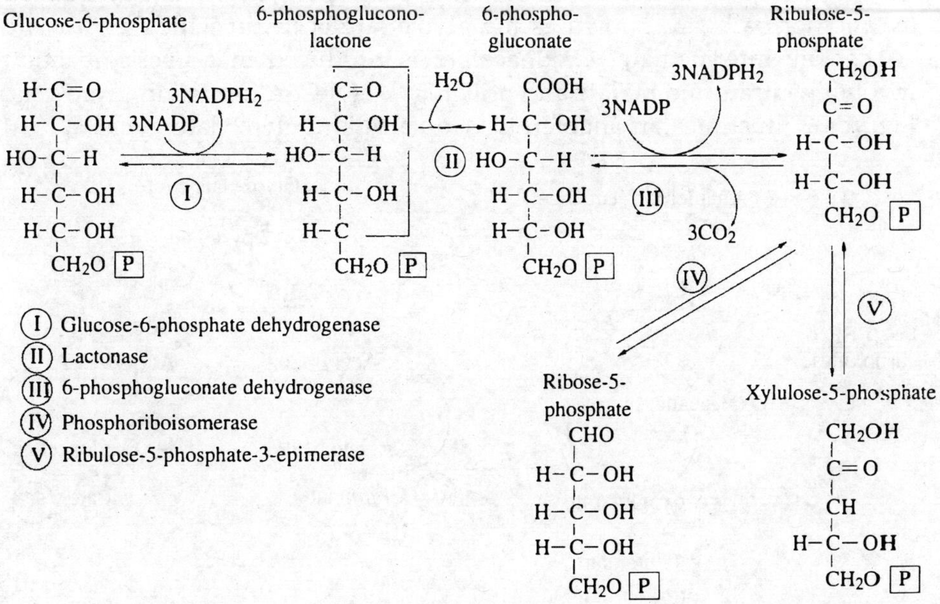

Fig. 31 Pentose phosphate pathway (Starting from glucose-6-phosphate).

In *Thiobacillus novellus* and *Brucella abortus* the non-oxidative reactions of the pentose phosphate pathway are used for the oxidation of glucose to CO_2, as they do not possess some key enzymes of EMP or ED pathways. 3 molecules of CO_2 and one molecule of glyceraldehyde-3-phosphate are formed, with the reduction of 6 molecules of NADP to $NADPH_2$.

The number of ADPs phosphorylated per one oxygen atom is known as the **P:O ratio**. In mitochondria it is 3 for one $NADH_2$ as electron donor and 2 for $FADH_2$ as electron donor. For some bacteria the P:O ratio may reach 3 but for *E. coli* it is usually 2.

Anaplerotic sequences (replenishing sequences). The diversion of oxaloacetate for biosynthetic pathways disturbs the TCA cycle. In order to overcome this, organisms have developed an alternate pathway for oxaloacetate formation from phosphoenol pyruvate (which is an intermediate compound in glycolysis). This reaction is catalyzed by the enzyme PEP carboxylase.

$$\text{PEP} + CO_2 \xrightarrow{\text{PEP carboxylase}} \text{Oxaloacetate} + \text{Pi}$$

α-oxoglutarate is also diverted from the TCA cycle. This also causes reduction in the available oxaloacetate, which is made up by this anaplerotic reaction.

Glyoxylate cycle. Due to diversion of 2 or 3 carbon intermediates in the TCA cycle for biosynthetic purposes, a sufficient quantity of oxaloacetate would not be available for continuing the TCA cycle. It is replenished by another pathway, known as the **glyoxylate cycle**, in which glyoxylate is the key intermediate. **Oxaloacetate** is produced in a cyclic reaction involving **citrate** and **malate** and utilizing Acetyl-CoA formed in glycolysis. Figure 32 illustrates an anaplerotic sequence and glyoxylate cycle.

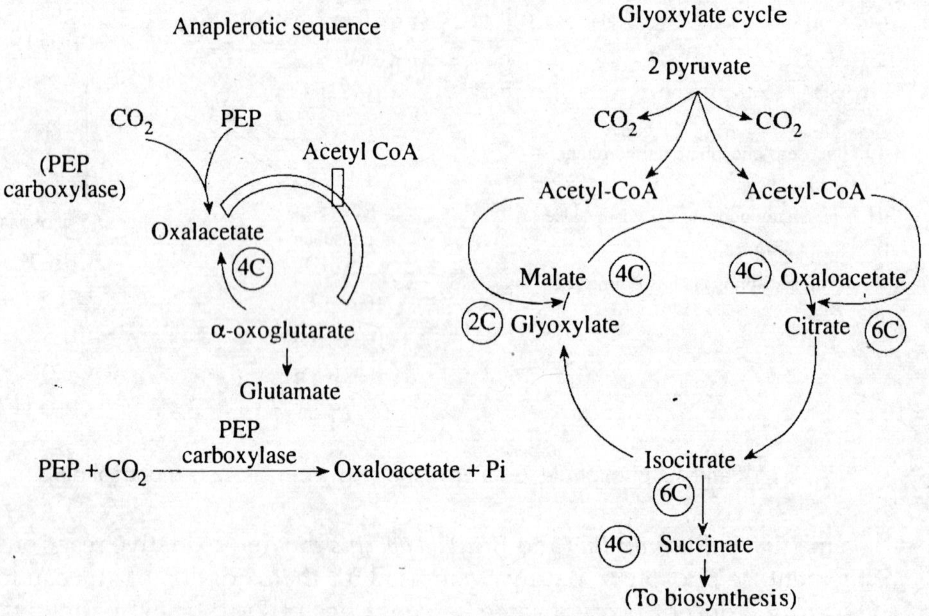

Fig. 32 Anaplerotic sequence and glyoxylate cycle.

Anaerobic Respiration

In the absence of oxygen, some organisms can use nitrate (NO_3^-), Sulphate (SO_4^-) and a few other compounds as terminal electron acceptors. This is described as **anaerobic respiration.** One such phenomenon, fairly widespread in nature, is **nitrogen respiration** or **denitrification** as it is called. The organisms capable of denitrification are known as **denitrifying bacteria.** *Thiobacillus denitrificans, Pseudomonas stutzeri, Alcaligenes faecalis, Bacillus licheniformis, Hyphomicrobium vulgare, Micrococcus denitrificans, Pseudomonas denitrificans* and *Achromobacter* sp.

A membrane-bound, molybdenum-containing enzyme, **nitrate reductase** reduces nitrate to nitrite.

$$NO_3^- + 2e^- + 2H^+ \longrightarrow NO_2 + H_2O$$

Nitrite reductase reduces NO_2 to nitric oxide (NO) and **nitric oxide reductase** reduces NO to nitrous oxide (N_2O).

$$NO_2^- + e^- + H^+ \longrightarrow NO + OH^-$$

$$2NO + 2e^- + 2H^+ \longrightarrow N_2 + H_2O$$

Since these compounds act as electron acceptors, **electron transport phosphorylation** occurs in the respiratory chain. As the molecular nitrogen that is formed escapes into the atmosphere, this process is also known as **dissimilative nitrate reduction**.

In some enteric bacteria, reduction of nitrate to molecular nitrogen may not occur. After the formation of nitrite, it is either excreted or further reduced to ammonia. Most non-denitrifying bacteria can also grow when nitrate is provided as the nitrogen source. The nitrate is reduced to ammonia by soluble enzyme systems and the process is known as **assimilatory nitrate reduction.** The ammonia thus formed is used for biosynthesis of various nitrogen-containing compounds.

The **sulphur-reducing bacteria** are strict anaerobes utilizing sulphate and other oxidized forms of sulphur, and a few facultative anaerobes which use elemental sulphur as electron acceptor (e.g. *Campilobacter*). *Desulfovibrio, Desulfomonas, Desulfococcus, Desulfobacter, Desulfosarcina* and *Desulfolobus* can utilize sulphate, sulphite or thiosulphate as electron acceptors. *Desulfuromonas* cannot use those compounds but can use elemental sulphur and some organic compounds like cystine and fumarate. The end product is usually **hydrogen sulphide**. The probable path is:

$$SO_4^{2-} + 8H \longrightarrow S^{2-} + 4H_2O \longrightarrow HSO_3^- \longrightarrow S_3O_3^{2-} \longrightarrow S_2O_3^{2-} \longrightarrow H_2S$$

(Sulphate + Adenine ⟶ Adenophosphosulphate ⟶ bisulphate ⟶ trithionite ⟶ thiosulphate ⟶ hydrogen sulphide)

Sulphide is quite toxic and its formation in water sometimes causes mortality of fish and other aquatic life. In waterlogged soils plants are also damaged. Corrosion of metal pipes also occurs by the activity of sulphur reducing bacteria.

Some organisms such as enteric bacteria are facultative anaerobes showing both oxidative and fermentative metabolism, while some are obligate anaerobes that are killed when exposed to oxygen. This toxic effect of oxygen is due to the fact that during electron transport, with oxygen as terminal electron acceptor, H_2O_2 and a superoxide radical (O_2^-) are formed. The enzymes **catalase** and **superoxide dismutase** convert them into water, thus eliminating the toxicity

$$O_2^- + O_2^- + 2H \xrightarrow{\text{Superoxide dismutase}} H_2O_2 + O_2$$

$$2H_2O_2 \xrightarrow{\text{Catalase}} 2H_2O + O_2$$

The superoxide dismutase is absent in obligate anaerobes, hence they are killed in the presence of oxygen.

BACTERIAL FERMENTATIONS

Several bacteria occur in environments which lack free oxygen. They obtain their energy through various fermentation processes. In most fermentation processes, ATP is produced at substrate level phosphorylation, though electron transport phosphorylation occurs in some instances.

Ethanol fermentation by yeast. Anaerobic breakdown of glucose and other hexoses by *Saccharomyces* species yields **ethanol** in industrial fermentation. The EMP pathway is employed for glucose breakdown. **Pyruvate** that is formed is decarboxylated to acetaldehyde by the enzyme pyruvate decarboxylase. CO_2 is released in the process. Acetaldehyde is reduced to **ethanol** by alcohol dehydrogenase. $NADH_2$ which is formed during glycolysis is oxidized in this process. Thus, the energy released in this is only 2 molecules of ATP by substrate level phosphorylation in glycolysis (Fig. 26). Hence, yeasts consume 3-4 times more glucose when transferred from aerobic to anaerobic conditions. The change from anaerobic to aerobic conditions results in lower glucose consumption and stoppage of alcohol formation. This phenomenon is known as the **Pasteur effect**. Yeasts are not regarded as typical facultative anaerobes because their growth stops after a few generations under anaerobic conditions. There are some facultative anaerobes in bacteria which carry out ethanol fermentation. *Zymomonas* species degrade glucose by the ED pathway, producing 2 moles of ethanol and 2 moles of CO_2 from one mole of glucose. *Erwinia amylovora* and *Sarcina ventriculi* also ferment glucose by the alcohol dehydrogenase system. They also employ the EMP pathway. Figure 33 illustrates the pathway of ethanol fermentation by yeast.

Some enteric bacteria, lactic acid bacteria and *Clostridia* also produce ethanol but they use a different pathway. **Acetaldehyde** is produced from acetyl-CoA and it is reduced to **ethanol** by the enzyme *acetaldehyde dehydrogenase*.

Alcoholic fermentation is the essential step in the **leavening** of dough in bread making. Moistening the flour releases enzymes which hydrolyze starch to sugar. Yeast cells feed on the released sugar. Formation of CO_2 and flavours by the yeast *Saccharomyces cerevisiae* improves the baking quality of dough.

Beer making from malted grain and **wine making** from grape juice are the result of fermentation by yeast. While selected cultures of *S. cerevisiae* are used in beer making, natural yeast flora on grapes is the usual source for fermentation, imparting characteristic flavours to different wines. In Japan and other Asian countries rice grain is also used for fermentation. Rice starch is first hydrolyzed by amylases from *Aspergillus oryzae* and then fermented. Hard liquors (those with higher ethanol content) such as brandy and whisky are obtained by distillation of fermented products.

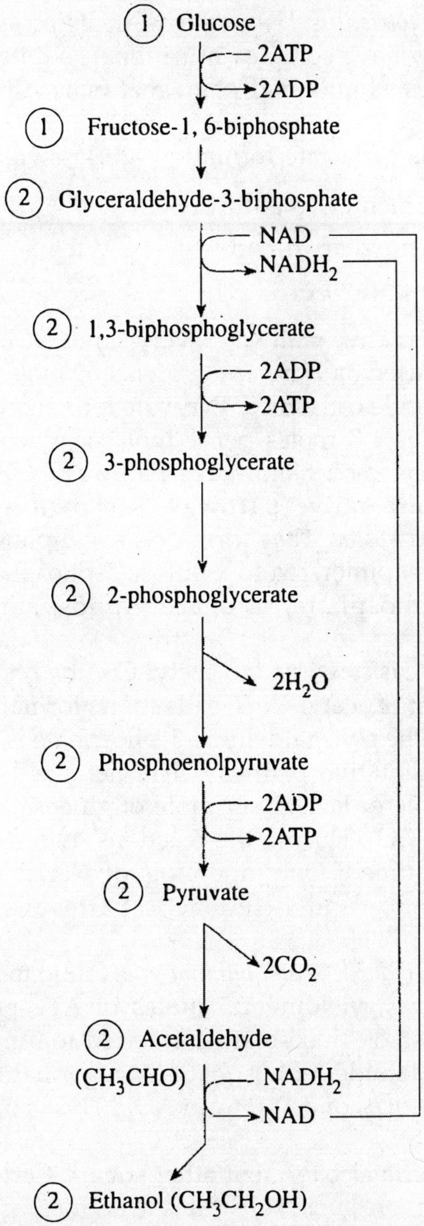

Fig. 33 Pathway of ethanol formation by yeast.

Contamination of fermenting juices by undesirable bacteria sometimes occurs. This was the reason for Louis Pasteur to take up the study of fermentations, and his work has been the most important aspect in the early development of microbiology.

Lactic acid fermentation. Several bacteria, known as lactic acid bacteria, form large quantities of lactic acid. Species of several genera such as

Lactobacillus, Sporolactobacillus, Leuconostoc, Streptococcus and *Bifidobacterium* ferment sugars. They have complex nutritional requirements and occur in such nutrient sources as milk and plant materials. They also occur in the intestines of animals.

Three pathways of lactate formation are known.

- The homofermentative pathway
- The heterofermentative pathway
- The bifidum pathway

The **homofermentative pathway** is very similar to ethanol fermentation. Pyruvate is directly used as H acceptor. Hence, 2 moles of ATP are formed for each glucose molecule consumed. Pyruvate formation is by EMP pathway. The yield of lactate is 1.8 moles per 1 mole of glucose. Most species of *Lactobacillus* and *Streptococcus* show this pathway.

The **heterofermentative pathway** is shown by some species of *Lactobacillus* and *Leuconostoc*. They form ribulose-5-phosphate as in pentose phosphate cycle. It is epimerized to xylulose-5-phosphate, cleaved to form glyceraldehyde-3-phosphate, and acetylphosphate by the enzyme phosphoketolase.

Acetylphosphate is converted to acetyl-CoA by phosphotransacetalase. Subsequent reduction of acetyl-CoA leads to the formation of acetaldehyde and finally ethanol. The glyceraldehyde-3-phosphate is converted to lactate as in the homofermentative pathway. The net yield in this pathway is production of 0.8 mole of lactate per mole of glucose, in addition to some acetate, ethanol and CO_2. During the reactions, two molecules of $NADH_2$ are formed and consumed. Only one mole of ATP is produced per mole of glucose. The homo- and heterolactic pathways are illustrated in Fig. 34.

The bacterium *Bifidobacterium bifidum* was found to break down glucose by a different pathway, yielding 2.5 moles of ATP per mole of glucose consumed. Two moles of glucose are converted to 3 moles of acetate and 2 moles of glyceraldehyde-3-phosphate. The formation of acetate from acetylphosphate produces one ATP from ADP. This is brought about by the enzyme acetatekinase.

Lactic acid bacteria also ferment other sugars. Certain variations occur during their growth on such substrates (fructose, galactose, mannose, saccharose, lactose etc).

Some lactobacilli are responsible for imparting the characteristic flavour to butter by forming acetoin and diacetyl. Citrate occurring in milk is the preferred substrate for these organisms (*Streptococcus cremoris* and *Leuconostoc cremoris*).

The production of lactic acid is useful for preservation of food as the pH of the medium is lowered, thus preventing the growth of other organisms. Many lactic acid bacteria occur in milk and cause curdling and development of characteristic flavours. Curdling is caused by the coagulation of milk proteins by lactic acid.

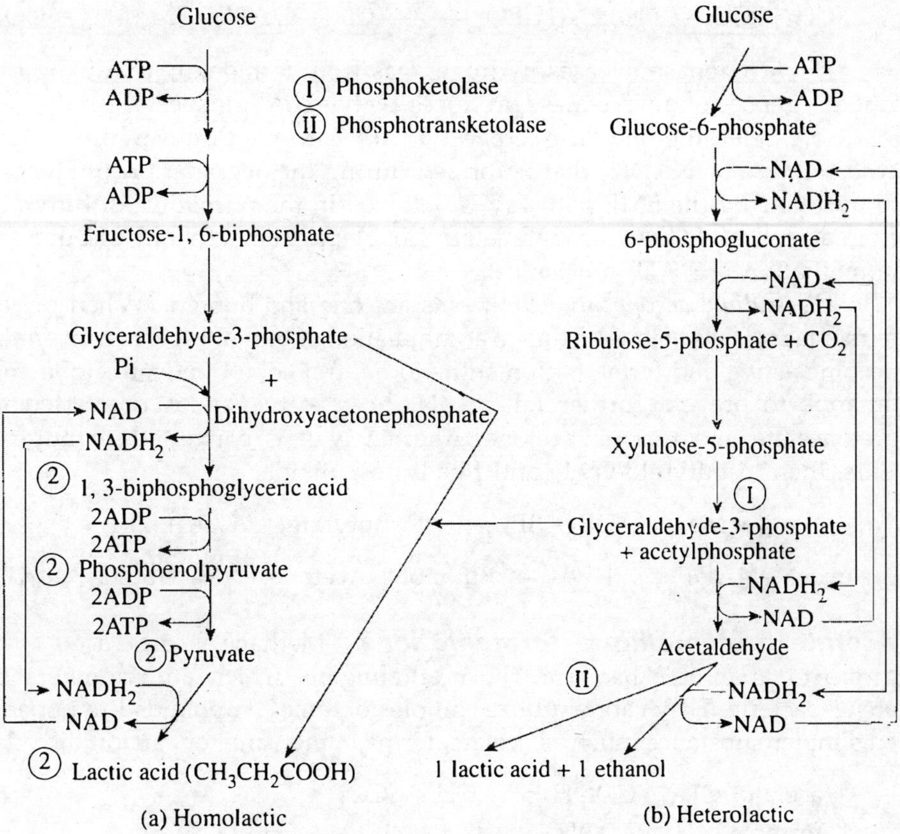

Fig. 34 Homolactic and heterolactic fermentations.

In cheese making, milk is curdled, excess water is drained away and the solids are subjected to the action of certain bacteria and fungi, in a process called ripening (for curdling of milk an enzyme *rennin* derived from the stomach of calves, is also sometimes added).

Lactic acid bacteria are also used in the preservation of some plant materials such as pickled olives and sauerkraut (pickled cabbage).

Silage is prepared by spraying animal fodder with lactobacilli so that its preservation and nutrient status are enhanced.

Butyrate and butanol-acetone fermentations. Butyrate is formed by obligate anaerobes *Clostridium, Butyrivibrio, Eubacterium limosum* and *Fusobacterium nucleatum*. The mechanism of butyrate formation has been thoroughly studied in *C. kluyveri* and *C. pasteurianum*. The discovery of ferredoxin in the latter clarified the mechanism. Ferredoxin is an iron-sulphur protein with low redox potential. (E'0–0.41 v).

In these bacteria, pyruvate is formed by the EMP pathway. Pyruvate is cleaved to form acetylphosphate releasing hydrogen gas and reducing ferredoxin. This reaction is known as a phosphoroclastic reaction (In other organisms NAD is reduced).

$$CH_3—CO—COOH + H_3PO_4 \longrightarrow CH_3—COOPO_3H_3 + H_2 + CO_2$$

A hydrogenase releases hydrogen gas from ferredoxin, oxidizing the latter. A phosphotransacetalase produces **acetyl-CoA** from acetyl phosphate and CoA, releasing one Pi. **Acetoacetyl CoA** is formed from two moles of acetyl CoA. The $NADH_2$ that is formed during the degradation of glucose to pyruvate by the EMP pathway is oxidized in the formation of **butyrate** from acetoacetyl CoA, through a series of steps. The net yield of butyrate fermentation is 3 ATP molecules.

Clostridium acetylobutylicum forms acetone and butanol. When pH of fermentation falls below 4.5 due to accumulation of butyrate, certain enzymes become active and fermentation shifts to neutral compounds, acetone and butanol, to prevent further fall in pH. Acetoacetyl-CoA is converted to acetoacetate and then to acetone. Alternately it is converted to butyryl CoA, then to butyraldehyde and finally to butanol.

$$\text{Glucose} + 3\text{ADP} + 3\text{Pi} \longrightarrow \text{Butyrate} + 2CO_2 + 2H_2$$

$$2\text{ glucose} + 4\text{ADP} + 4\text{Pi} \longrightarrow \text{Butanol} + \text{Acetone} + 4H_2 + 5CO_2 + 4\text{ATP}$$

Acetate and methane fermentations. Methane is produced by a group of anaerobic bacteria, the **methanogens** which are members of archaebacteria. They cannot utilize complex organic compounds but depend on simpler substances such as acetate, formate, methanol or carbon dioxide.

$$\text{(acetate) } CH_3—COOH \longrightarrow CH_4 + CO_2$$
$$\text{(formate) } 4H—COOH \longrightarrow CH_4 + 3CO_2 + 2H_2O$$
$$\text{(methanol) } 4CH_3—OH \longrightarrow 3CH_4 + CO_2 + 2H_2O$$
$$\text{(carbon dioxide) } CO_2 + 4H_2 \longrightarrow CH_4 + 2H_2O$$

Since the methanogenic bacteria utilize CO_2 and H_2, they are regarded as *chemolithotrophs.* However, they occur along with fermentative bacteria under strict anaerobic conditions and are effective in trapping the CO_2 and H_2 evolved in those systems, resulting in methane formation. The organisms are species of *Methanococcus, Methanobacterium, Methanosarcina* and *Methanospirillum*. As methane is the most reduced compound, it is an important energy source. Anaerobic digesters are used to produce methane from organic wastes and to use it as fuel. The energy-giving material in *gobar gas* is methane.

The vinegar-producing bacteria are aerobic, Gram-negative bacteria, generally known as acetic acid bacteria. Vinegar is produced by converting ethanol to acetic acid. Fermented fruit juices are the usual source of ethanol. Industrial acetic acid (distilled vinegar) is produced from pure ethanol and water. Vinegars from different sources have characteristic flavours. The acetic acid bacteria are species of *Acetobacter* and *Gluconobacter*. They do not, however, oxidize glucose completely but form acetic acid while utilizing oxygen as terminal electron acceptor. ATP is generated in this electron transfer to oxygen.

$$CH_3{-}CH_2OH + PQQ \xrightarrow[\text{Dehydrogenase}]{\text{Alcohol}} CH_3CHO + PQQH_2$$

$$CH_3{-}CHO + PQQ \xrightarrow[\text{Dehydrogenase}]{\text{Aldehyde}} CH_3COOH + PQQH_2$$

(PQQ = methaxatin – a coenzyme)

The reduced $PQQH_2$ is oxidized in the electron carrier respiratory chain, resulting in ATP formation.

Some species of *Clostridium* (*C. thermoaceticum, C. formicoaceticum*) convert hexoses to acetate. Glucose is degraded to pyruvate through the EMP pathway. Further reactions produce two acetate molecules. *C. formicoaceticum* uses fructose.

$$C_6H_{12}O_6 \longrightarrow 3CH_3COOH$$

These acetogenic bacteria are anaerobic and differ from vinegar-producing bacteria which are aerobic.

BIOSYNTHESIS OF CELLULAR COMPOUNDS

The ATP generated in catabolic processes is utilized to a large extent for the synthesis of macromolecules of the cell. It is estimated that 56% is used up for the synthesis of amino acids and their polymerization into proteins, 20% is used up for nucleic acid synthesis and turnover of messenger RNA (which occurs rapidly), about 7% for polysaccharide and lipid synthesis, and the remaining for transport of salts and other purposes.

Amino acids. The amino acids are synthesized mostly from products of glycolysis and the TCA cycle. The precursors of various amino acids found in different proteins are listed in the Table 11. Their structures are illustrated in Fig. 35.

Table 11 Precursors of Common Amino Acids

Precursors	*Amino acids*
Pyruvate	– Alanine, Valine, Leucine
Oxaloacetate	– Aspartate, Asparagine, Methionine, Lysine, Threonine, Isoleucine
α-oxoglutarate	– Glutamate, Glutamine, Arginine, Proline
3-phosphoglycerate	– Serine, Glycrine, Cysteine
Phosphoenolpyruvate	– Phenylalanine
Erythrose-4-phosphate	– Tyrosine, Tryptophan
5-phosphoribosyl-1-pyrophosphate + ATP	– Histidine

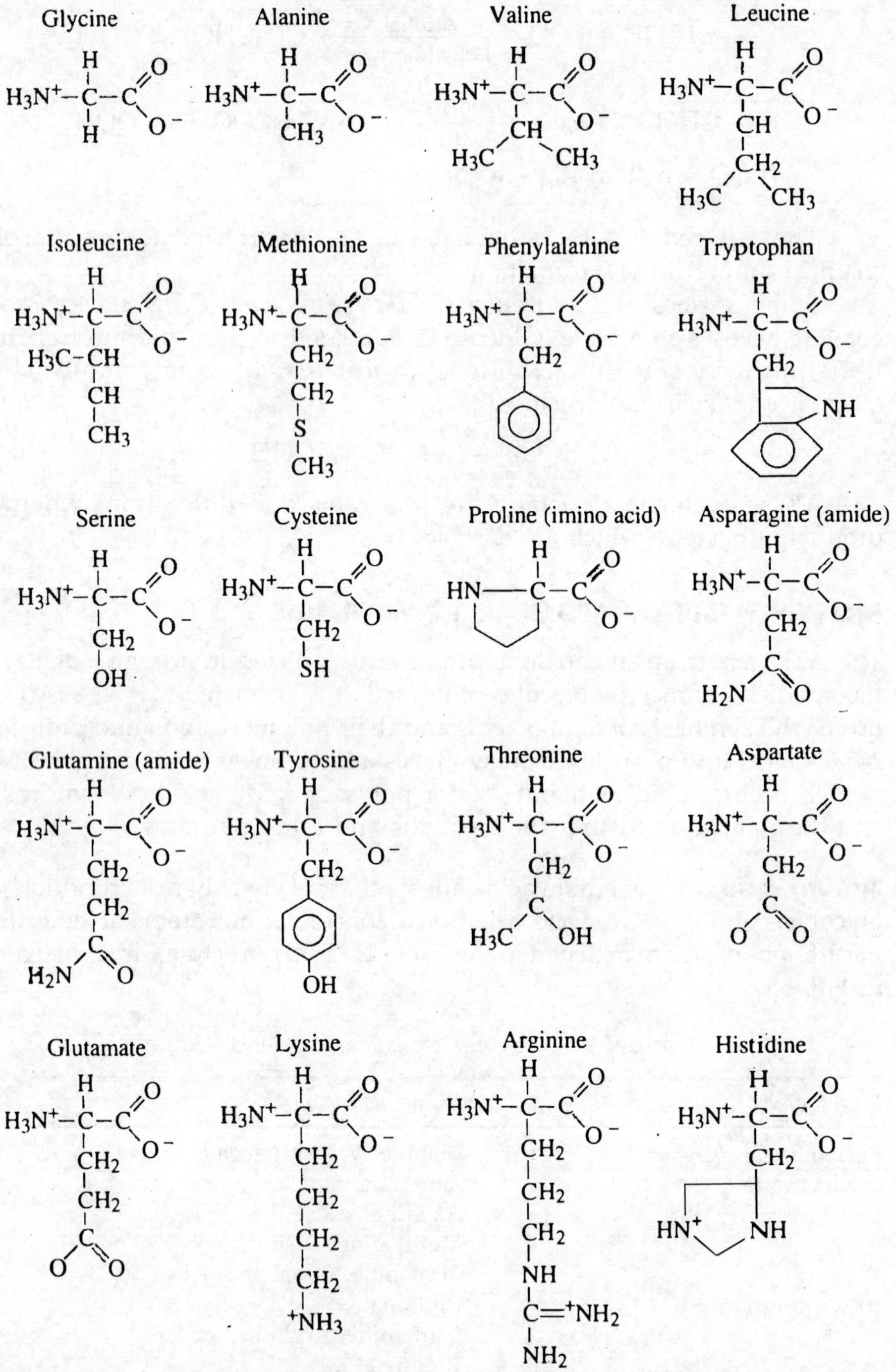

Fig. 35 Structures of common amino acids.

Ammonia assimilation. When sufficient supply of ammonia is present, reductive amination of α-oxoglutarate (generated in the TCA cycle) results in the formation of L-glutamate. This reaction is catalyzed by the enzyme

L-glutamate dehydrogenase. The amino group of L-glutamate can be transferred to other L-oxoacids by transaminases.

Glutamine is produced from glutamate in a reaction brought about by L-glutamine synthetase. Glutamine is important as it acts as a NH_2 donor for the synthesis of purines, pyrimidines and some amino acids. When ammonia concentration is low, it has been shown that a combination of glutamine synthetase (GS) and glutamate synthetase (Glutamine-α-oxoglutarate aminotransferase, or GOGAT) catalyse the formation of glutamate (Fig. 42).

Assimilative nitrate reduction. When nitrate is provided as the nitrogen source, its assimilation is by reduction to ammonia before it is incorporated into the cellular compound. This is called **assimilative nitrogen reduction** and is carried out by soluble enzymes which are ammonia depressed. This is in contrast to the membrane-bound nitrate reductase which is active in denitrification (**dissimilative nitrate reduction**). Assimilative nitrate reduction occurs in all plants, most fungi and many bacteria, whereas dissimilative nitrate reduction is confined to some bacteria.

Assimilative sulphate reduction. Sulphate is the principal source of sulphur for most bacteria. It is reduced to H_2S before being incorporated into cellular compounds. Sulphate is actively transported into the cell. Since it is a fairly stable ion it has to be activated before combining with other compounds. The enzyme ATPsulphurylase catalyses the attachment of the sulphate ion to ATP resulting in the formation of adenosine phospho-sulphate (APS). In assimilative reduction, the next step is the addition of another P to form phosphoadenosine phosphosulphate (PAPS). This is brought about by the enzyme APS phosphokinase. The PAPS combines with thioredoxin (RSH) to release sulphite (SO_3^{2-}). This is catalyzed by PAPS reductase. The enzyme sulphate reductase then reduces sulphite to H_2S, trasferring 6 electrons from 3 $NADPH_2$. Hydrogen sulphide (which is a toxic compound), is immediately incorporated into O-acetylserine. (In dissimilatory sulphate reduction, the H_2S is excreted.)

Biosynthesis of Nucleotides. Nucleotides are made up of purine or pyrimidine bases, the sugars ribose or deoxyribose and one or more phosphate groups (Fig. 36). The attachment of a purine or pyrimidine base to the sugar forms a nucleoside. Esterification of a nucleoside with phosphoric acid yields nucleotide.

The ribonucleotides containing pyrimidine bases are formed from the precursors **carbamylphosphate** and **aspartate**. The enzyme aspartate transcarbamylase condenses them to yield 4, 5-dihydroorotate. Dehydrogenation of this compound yields orotate. It is linked to phosphoribosyl-1-pyrophosphate (PRPP) yielding orotidine monophosphate. This is decarboxylated to form **uredine monophosphate** (UMP). UMP is converted to uredine triphosphate (UTP) by kinase reactions. Cytosine triphosphate (CTP) is generated from UTP.

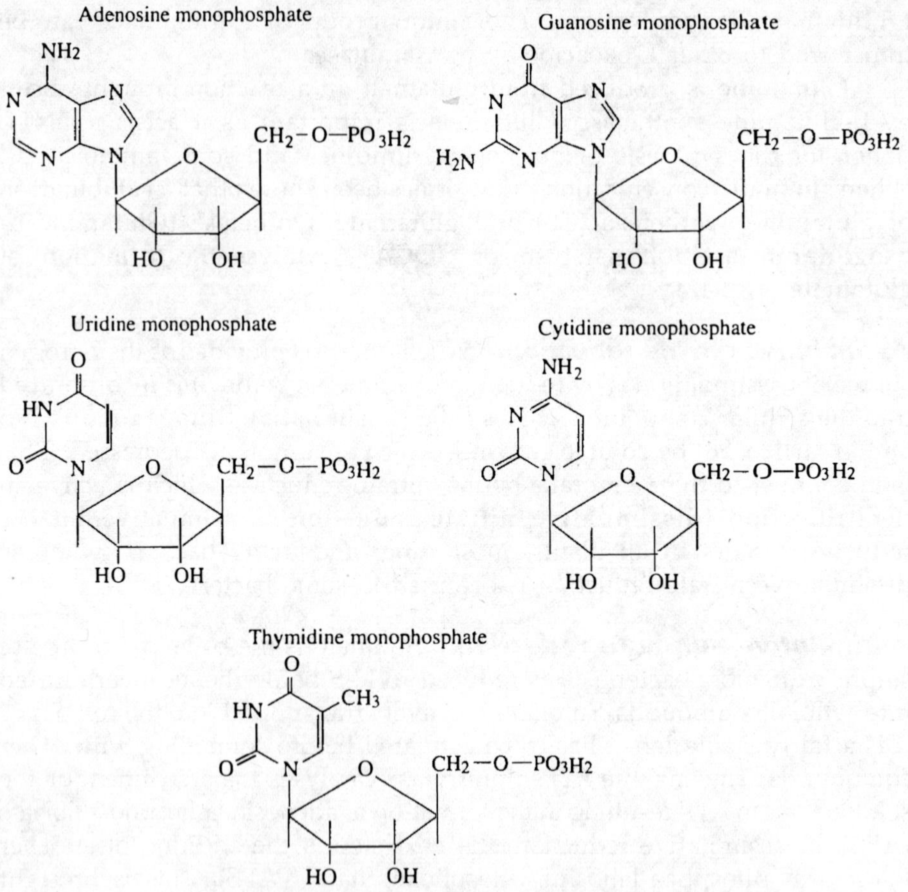

Fig. 36 Structure of nucleotides.

The formation of purine nucleotides is more complicated. From PRPP, 5-aminoimidazole ribonucleotide is formed. Bicarbonate, aspartate and formyl-H_4-folate are involved in the formation of the purine ring. Further reactions result in the formation of adenosine monophosphate (AMP) and guanidine monophosphate (GMP). Subsequent reactions brought about by kinases result in the formation of adenosine triphosphate (ATP) and guanidine triphosphate (GTP). Deoxyribonucleotides are formed by reduction of ribonucleotides at the pyrophosphate level.

Biosynthesis of Lipids. Lipids are fatty acids esterified with glycerol. They are found in membrane systems of bacteria although they are not stored as reserve materials as is the case in higher organisms. Phospholipids and glycolipids are the major constituents of membranes. Phospholipids contain other moieties like choline, inositol, serine and ethanolamine. Glycolipids contain sugars like mannitol, glucose, galactose and oligosaccharides.

Fatty acid synthesis. Most of the lipids contain 16 or 18 carbon fatty acids. The fatty acids exhibit an interesting property in that they have both hydrophilic and hydrophobic regions. Palmitate (16 carbon atom) is a saturated fatty acid, stearate (18 carbon) is also a saturated one while oleate (18 carbon) is a monounsaturated fatty acid.

An acyl carrier protein (ACP) resembling CoA is acetylated by acetyl-CoA to form acetyl-ACP. This compound acts as a primer in fatty acid synthesis. Two C_2 units are added to acetyl ACP to form malonyl ACP. This is carried out by the enzymes acetyl CoA carboxylase and malonyl-transacelase. The action of four enzymes is required for the extension of the acetyl-ACP by two carbon atoms, as shown below:

$$\text{Acetyl-ACP} + \text{malonyl-ACP} \xrightarrow[\text{Synthetase}]{\text{3-ketoacyl-ACP}} \text{Acetoacetyl-ACP} + CO_2 + \text{ACP}$$

$$\text{Acetoacetyl-ACP} + NADPH_2 \xrightarrow[\text{ACP reductase}]{\text{3-ketoacyl}} \beta\text{-hydroxybutyryl-ACP} + \text{NADP}$$

$$\beta\text{-hydroxybutyryl-ACP} \xrightarrow[\text{Dehydratase}]{\beta\text{-hydroxyacyl}} \text{Crotonyl ACP} + H_2O \longrightarrow \text{ACP}$$

$$\text{Crotonyl-ACP} + NADPH_2 \xrightarrow[\text{Reductase}]{\text{Enoyl ACP}} \text{Butyryl-ACP} + \text{NADP}$$

These enzymes are relatively unspecific and catalyze further addition of C_2 units until a 16- or 18-carbon fatty acid is formed.

Phospholipid formation. Phospholipids are formed from phosphatidic acids. Glycerol-3-phosphate and acyl-ACP are combined to form **phosphatidic acid**, with the release of ACP. Phosphatidic acid is combined with CTP (cytosine triphosphate) to form **CDP-diacylglycerol**, releasing inorganic phosphate, Pi. The CDP-diacylglycerol combines with an amino acid to form a phospholipid. For instance, it combines with serine to form phosphatidylcholine. The chemical properties of phospholipids make them the ideal components of membranes.

Formation of Polymers

Carbohydrates. The intermediate products of the EMP and the oxidative pentose phosphate cycles like glucose-6-phosphate, fructose-6-phosphate, fructose-1,6-biphosphate, ribulose-5-phosphate, xylulose-5-phosphate and ribose-5-phosphate are utilized for the synthesis of carbohydrate polymers. Glucose-6-phosphate is first converted to glucose-1-phosphate by the enzyme **phosphoglucomutase.** This combines with UTP to form **UDP-glucose**, the reaction catalyzed by UDP-glucose **pyrophosphorylase.** An **epimerase** with NAD converts UDP-glucose to **UDP-galactose.** This latter compound is the precursor of lipopolysaccharide outer membrane of *E. coli*.

The **hexoseamines**, which are the other precursors of the cell wall originate from fructose-6-phosphate. UDP-N-acetyl-muramic acid, mother component of the cell wall is also synthesized from the same source. Fructose-6-phosphate takes up NH_2 from glutamine to form glucosoamine-6-

phosphate and glutamic acid, the reaction being catalyzed by **amino-transferase.** Glucosamine-6-phosphate reacts with acetyl-CoA to form N-acetyl-glucosamine-1-phosphate. This is linked with UDP to form UDP-N-acetylmuramic acid. The peptidoglycan is synthesized by reaction of this compound with amino acids.

The capsule of bacteria is made up of polysaccharides and glyco-proteins. The term **glycocalyx** is used to designate the slime layer or capsule. The filaments of the glycocalyx are formed from sugars, sugar alcohols and amino sugars. They are rigid or flexible depending on the constituents.

Polyhydroxybutyrate is a typical storage material in many bacteria. It is derived from acetyl-CoA. **Polyphosphate** is another storage material and so is **glycogen**. It is a highly branched polymer consisting of glucose monomers linked together by α-1, 4 or α-1, 6 linkages. Glucose is first activated by the formation of ADP-glucose.

$$\text{Glucose-1-phosphate} + \text{ATP} \longrightarrow \text{ADP-glucose} + \text{Pi}$$

This is catalyzed by the enzyme **ADP-glucose pyrophosphorylase.** The glucose moiety is then transferred to an oligosaccharide primer. The enzyme catalyzing this is **glycogen synthetase.** Further elongation of the molecule occurs in similar transfers.

Plasma membrane and outer membrane. The plasma membrane is made up of **phospholipid molecules,** such as phosphatidyl choline, associated with one another to form a double layer. Membrane proteins and lipids are associated with the phospholipids.

The outer membrane layer in Gram-negative bacteria is made up of 5% lipopolysaccharide, 35% phospholipids and 15% protein. The lipopolysaccharide consists of fatty acids, notably 3-hydroxytetradecanoic acid, linked to glucosamine disaccharide and a core oligosaccharide consisting of several carbohydrate monomers.

Synthesis of DNA and RNA molecules. The nucleotide bases, **adenine, guanine,** and **cytosine** are present in both DNA and RNA, while **thymine** is present only in DNA and **uracil** only in RNA The nucleic acids are **polynucleotides** in which the nucleotides are **covalently bonded** to one another. The sequence of nucleotides is specific for any given polynucleotide. Sugar and phosphate moieties form the backbone. The phosphate linkage is known as a **diester bond** because it is connected to two sugars by ester linkage. The five nucleotides are designated by their first letters-A (adenine), G (guanine), T (thymine), C (cytosine) and U (uracil).

DNA (Deoxyribonucleic acid). According to the Watson-Crick model the DNA molecule is a **double helix** formed by two strands of poly-nucleotides held together by hydrogen bonds. The strands run in opposite directions. The 5' or phosphate end and the 3' or OH end of the nucleotides run in opposite directions in the two strands as shown in Fig. 37. Effective

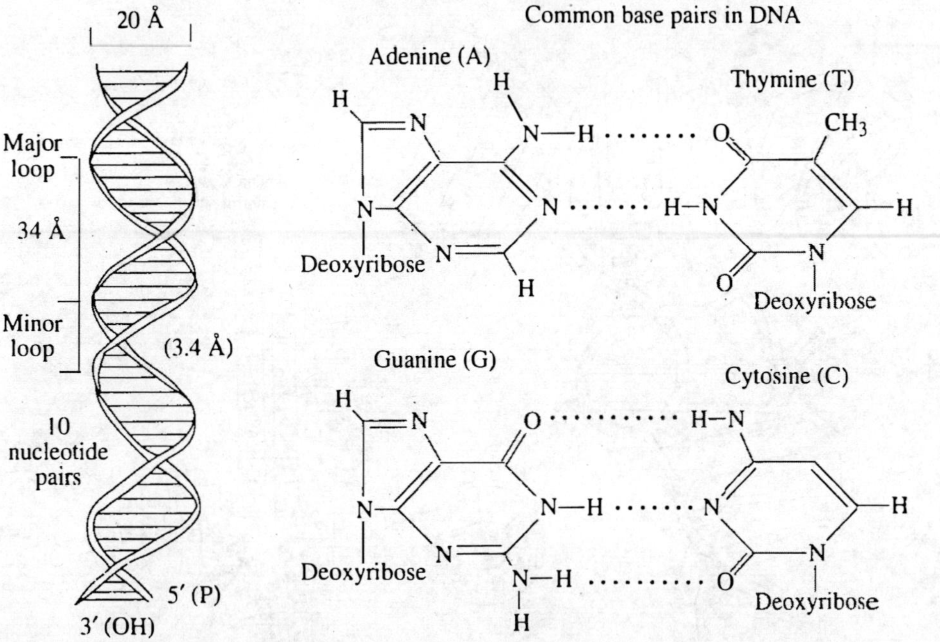

Fig. 37 DNA double helix and bonding of nucleotides.

hydrogen bonds can form between adenine and thymine and between guanine and cytosine. The mechanism of separation of the two strands makes it possible to duplicate the information. The two strands separate and new strands are formed along the parental strands. This process is catalyzed by **DNA polymerases** and the process is known as **replication**. This is a **semi-conservative method** as the two parental strands remain intact, while the other two strands are newly formed.

RNA (Ribonucleic acid). The formation of the RNA molecule is similar to that of the DNA molecule. It is catalyzed by **DNA-dependent RNA polymerase**. The process is known as **transcription**. Only the required part of the DNA codes the transcription depending on the signals it receives. Instead of thymine, uracil pairs with cytosine in RNA. Different types of RNA are produced for different purposes. A **messenger RNA (mRNA)** carries information for incorporation of amino acids to form polypeptide chains. It has a very short life, and splits up into monomers as soon as its function is over, thus facilitating the formation of new RNA molecules. The second type is the **ribosomal RNA** (rRNA) which forms the structural component of ribosomes. The third type is the **transfer RNA (tRNA)** which transfers amino acids to the polypeptide chain. This process is known as **translation.** This function is the basis for protein synthesis. Figure 38 shows the 3-dimensional structure and configuration of a tRNA molecule.

Protein Synthesis

Proteins are made up of one or more polypeptide chains. The number of

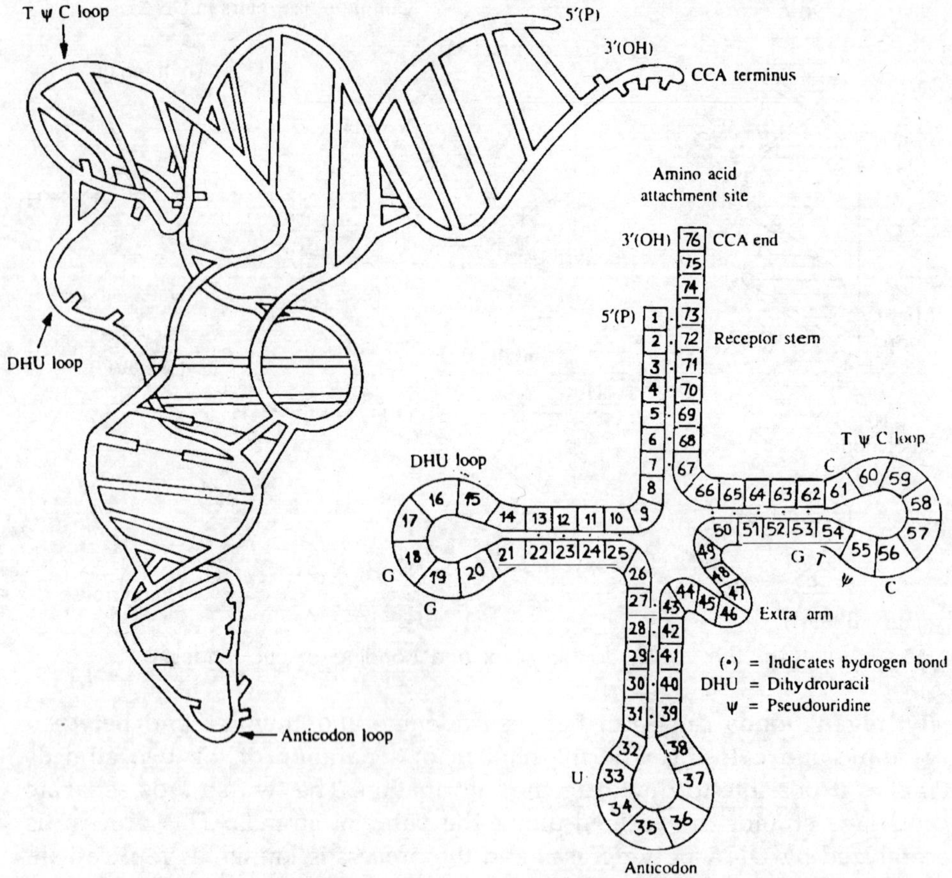

Fig. 38 Diagram of yeast tRNA^{phe} (three dimensional view and clover-leaf diagram).

amino acids, the sequence of their occurrence and the manner in which they are arranged, results in the occurrence of a large number of protein types. The discovery of the **triplet code** for amino acids facilitated the study of **protein synthesis**. Messenger RNA carries the triplet codes for the sequence of amino acids to be attached. The mRNA is attached to ribosomes which catalyze the formation of **peptide bonds**. There are specific sites (**anticodon loops**) on transfer RNA molecules linked to corresponding amino acids by the **aminoacyl-tRNA** synthetases, which are specific to the amino acids. The tRNA anticodon region recognizes the code on the mRNA. The formation of peptide bonds between the sequentially added amino acids results in the formation of **polypeptide**. For each peptide bond formation, an energy equivalent of 4 ATP molecules is expended.

Protein molecules often occur in folded fashion. The linear sequence of a protein is known as its **primary structure**. The polypeptide chain gets folded in a specific manner forming its **secondary structure**. Formation of hydrogen bonds between some amino acids leads to this folding. The secondary structure also folds back on itself in a particular way to form a stable molecule. This is the **tertiary structure**, resulting in a three-dimensional shape. In proteins having more than one polypeptide chain, a **quaternary structure** is formed, with subunits having a **tertiary structure**. The folding of the protein molecule is important because active sites where enzyme substrate complex is formed are positioned by this folding.

Utilization of organic substrates. Microorganisms have the ability to decompose a great variety of organic substrates to satisfy their energy and biosynthetic needs. When plants and animals die and reach soil or water, several substances such as organic acids, amino acids and other low molecular compounds as well as polymers like cellulose, starch, other polysaccharides, proteins, nucleic acids, aromatic compounds and hydrocarbons are attacked by microorganisms.

Degradation of polymers. Polymers in the environment are broken down to small molecules by the secretion of **exoenzymes**, before they are transported to the cell. Figure 39 illustrates the structure of glycogen and cellulose. Glycogen, which is a common storage carbohydrate, has a structure similar to that of starch but is smaller in length.

Starch and glycogen are degraded by **α-amylases** secreted by many bacteria. They are first cleaved into dextrines and then into glucose, maltose and other oligosaccharides. α-1, 4 linkages of straight chains are cleaved leaving the α-1, 6 linkages which occur at branch points.

Cellulose is the major component of plant cell walls and constitutes the largest organic source available to microbes. The large cellulose molecule is broken up in several steps. Several organisms secrete **extracellular cellulases** which cleave the molecule outside. They occur in soil, water and the rumen of herbivorus animals. The **cellulases** break down β-1, 4 linkages giving rise to **cellobiose** (disaccharide) and glucose. Both these molecules are taken up by the microorganisms. The enzymes that act on native cellulose are known as **C1 enzymes** and those that further cleave the resulting dextrins are called **C χ enzymes**. Cellobiase or **β-glucosidase** cleaves the β-1,4 linkage of cellobiose, releasing **glucose** molecules.

Cellulose-degrading bacteria include *Cytophaga*, *Cellulomonas*, *Bacteroides*, *Ruminococcus* and species of *Bacillus* and *Clostridium*. Many fungi are capable of degrading cellulose. Inside the cell, cellobiose is phosphorylated to glucose-1-phosphate and glucose by the enzyme *cellulobiose phosphorylase*. This reaction does not require ATP, and is therefore, economical. Maltose and sucrose are also phosphorylated in the same way by the related enzymes *maltose phosphorylase* and *sucrose phosphorylase*, respectively.

Pectic substances in plant cell walls are degraded by several fungi and bacteria (*Bacillus*, *Clostridium*, *Erwinia*, *Corynebacterium*, *Pseudomonas etc.*) Some soil bacteria and actinomycetes degrade **chitin.** The enzymes are known as **pectinases** and **chitinases**, respectively. Their action is extracellular. Pectins are hydrated polysaccharides containing galacturonic acid units

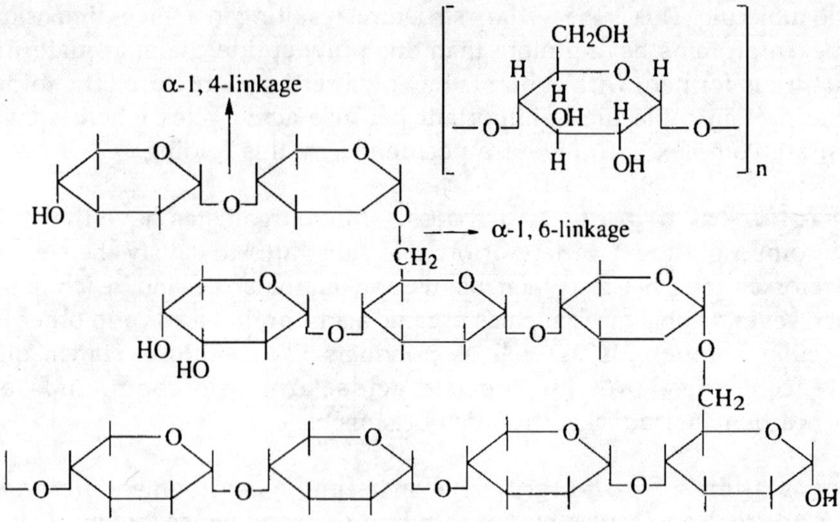

Structure of part of glycogen molecule

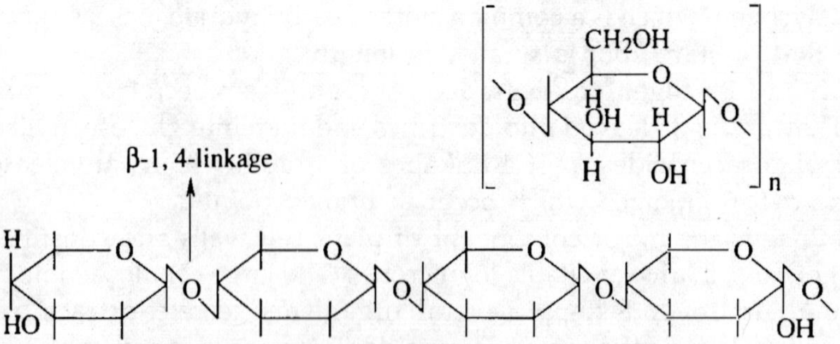

Structure of part of cellulose molecule

Fig. 39 Diagram showing partial structure of cellulose and glycogen molecules.

and also rhamnose to a small extent. Chitin is a polymer of N-acetylglucosamine and is a common constituent of fungal cell walls as also shells and nails of some animals.

Nucleic acids are degraded by ribonucleases (RNAses) (e.g. *Bacillus subtilis*), and deoxyribonucleases (DNAses) (e.g. *Staphylococcus aureus* and *Clostridium*).

Utilization of Amino Acids

Many bacteria produce **proteases** which break down proteins to yield constituent amino acids or small molecular weight peptides before they are taken up by the cells. Oxidative deamination converts the amino acids into the corresponding organic acids with the release of ammonia. The energy required is obtained either through cytochromes or through NAD-linked dehydrogenation, (which occurs in the deamination of alanine or glutamate).

$$R\text{—}\underset{\underset{NH_2}{|}}{CH}\text{—}COOH + \tfrac{1}{2}\,O_2 \longrightarrow R\text{—}CO\text{—}COOH + NH_3$$

$$CH_3\text{—}CH\text{—}COOH + NAD + H_2O \longrightarrow CH_3\text{—}CO\text{—}COOH + NADH_2 + NH_3 \text{ (pyruvate)}$$

Deamination of some amino acids (e.g. serine) is brought about directly through the action of dehydratases.

$$\underset{\underset{OH}{|}}{CH_2}\text{—}\underset{\underset{NH_2}{|}}{CH}\text{—}COOH \longrightarrow CH_3\text{—}CO\text{—}COOH + NH_3$$

Many bacteria grow well on nutrient media containing mainly amino acids because of their ability to use them both as carbon and nitrogen sources.

Utilization of Glyoxylate

Glyoxylate appears as a breakdown product of nucleic acids. It is introduced into a cyclic reaction yielding reducing power. This is known as the **dicarboxylic acid cycle** (glyoxylate cycle, as already explained).

Utilization of Aliphatic Hydrocarbons

Growth on hydrocarbons is always aerobic because oxygen is a primary element of organic substances. Two groups of organisms are involved, those that utilize methane, a C4 compound, and those that grow on other hydrocarbons. These compounds are insoluble in water and their movement into the cells involves special transport mechanisms. Some bacteria (e.g. *Pseudomonas fluorescence, Nocardia petroliphila*), yeasts (*Candida lipolytica*) and the fungus *Cephalosporium roseum* utilize them. Their study has attracted attention for two reasons: They develop in fuel injection systems and cause problems, particularly in aircraft, and, they are used for production of cellular **proteins** from petroleum wastes.

Utilization of Aromatic Compounds

Lignin is produced by plants in large quantities, forming part of secondary walls of plant cells. It is made up of **phenolic acids** which are aromatic compounds. Some of the amino acids and vitamins also have aromatic structure. Such substances are degraded by some bacteria and fungi. They are generally converted to catechol and protocatechuic acid. The aromatic ring is then cleaved, finally producing **succinate**. Degradation of phenylalanine and tyrosine, which are aromatic amino acids, results in the formation of homogentisic acid as an intermediate. Its degradation results in the formation of fumarate and acetoacetate. These reactions involve oxidation steps, hence only aerobic degradations occur.

METABOLISM OF AUTOTROPHS

Among bacteria, photosynthetic bacteria carry out anoxygenic photosynthesis while cyanobacteria carry out oxygenic photosynthesis. The former lack chlorophyll-*a*, whereas cyanobacteria and the eukaryotic algae possess chlorophyll-*a*. Recently, a new group of prokaryotic organisms, the prochlorophytes possessing chlorophyll-*a* have been discovered.

Photosynthetic bacteria cannot use hydrogen since they are anaerobes, hence depend on other hydrogen donors such as organic compounds, sulphur compounds and molecular hydrogen. Depending on structural differences and their sources of hydrogen, they are grouped under *Rhodospirillaceae* (purple non-sulphur bacteria) *Chromatiaceae* (purple sulphur bacteria) and **Chlorobiaceae** (green bacteria). Only cyclic photophosphorylation (**Photosystem I**) occurs in these organisms, whereas in cyanobacteria and eukaryotic algae as well as in higher plants, both non-cyclic and cyclic photophorylations occur, the light quanta being absorbed at 680 and 700 nm, wavelengths respectively (Photosystems I and II). CO_2 is used as the carbon source. The two photophosphorylation systems are illustrated in Fig. 40. When H_2S is present it is utilized as a source of electrons, resulting in the accumulation of elemental sulphur which is deposited either inside the cells (purple sulphur bacteria), or outside (green bacteria). When sulphide concentration is low, elemental sulphur is utilized, forming sulphate.

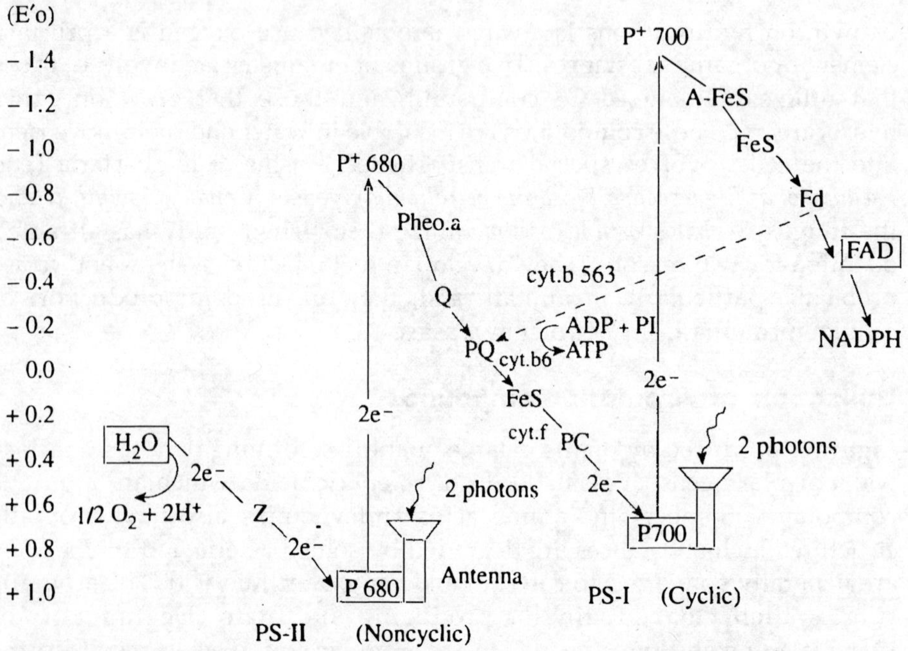

Pheo.a = Pheophytin; Q = Quinone (unknown); PQ = Plastoquinone; Cyt = Cytochromes; FeS = Ferrous-sulphur proteins; PC = Plastocyanine; Fd = Ferredoxin; Z = (unknown)

Fig. 40 Light reactions in photosynthesis.

$$CO_2 + H_2S \xrightarrow{\text{Light}} (CH_2O) + 2S + H_2O$$

$$CO_2 + 2S + 5H_2S \xrightarrow{\text{Light}} (CH_2O) + 2H_2SO_4$$

Rhodospirillum rubrum was shown to produce **hydrogen gas** when growing on organic and amino acids. This is of importance in our search for alternate energy sources, as hydrogen is the most efficient energy source.

$$CH_3\text{—}COOH + 2H_2O \xrightarrow{\text{Light}} 2CO_2 + 4H_2$$

Purple bacteria employ the **Calvin cycle** for CO_2 fixation. The mechanism in green sulphur bacteria is not well understood.

Chemolithotrops

Chemolithotrophs obtain energy by oxidizing inorganic compounds. Five groups are recognized, based on the type of compounds oxidized.

Nitrifying bacteria. The organisms that oxidize ammonia to nitrite, and nitrite to nitrate are together called **nitrifying bacteria**. These are **obligate autotrophs** and are inhibited under reducing conditions.

Oxidation of ammonia to nitrite is carried out by *Nitrosomonas, Nitrosospira, Nitrosococcus* and *Nitrosovibrio.* Those that oxidize nitrite to nitrate are *Nitrobacter, Nitrococcus* and *Nitrospira.* The first step in the oxidation of ammonia to nitrite is the formation of hydroxylamine as an intermediate, through a monoxygenate reaction requiring molecular oxygen. Further oxidation is through a hypothetical compound nitrohydroxyl, with cytochromes transferring electrons to oxygen.

$$NH_3 \xrightarrow{O_2 + H_2O} NH_2OH \rightarrow (NOH) \longrightarrow NO_2^-$$

Nitrite as well as hydroxylamine are toxic and it is of importance that they do not accumulate. But NO_2 is taken up by the second group of bacteria which further oxidize it to **nitrate.** Oxidation of nitrite to nitrate involves only electron transport, the oxygen source being water. Hence molecular oxygen is not directly needed for this step.

$$NO_2^- + H_2O \longrightarrow NO_3^- + 2H^+ + 2e^-$$

The activity of nitrifying bacteria is important for soil fertility as most plants use nitrate as nitrogen source. *Nitrosomonas* occurs in soil, water and sewage while *Nitrosospira* and *Nitrosovibrio* are purely soil organisms. Some species of *Nitrosococcus* occur in soil and some in marine habitats. *Nitrobacter* occurs in soil and water while *Nitrospira* and *Nitrococcus* are marine forms.

Sulphur oxidizing organisms. The compounds of sulphur occur in nature from a highly oxidized state of **sulphate** (+6) to the **hydrogen sulphide**

(–2). Organic forms include sulphur-containing amino acids (cystine, methionine) and a few other compounds such as thiamine, biotin and CoA. Oxidation of inorganic sulphur compounds is carried out by a number of microorganisms in soil and water.

Thiobacillus thiooxidans is strictly aerobic and grows in acidic soils. *T. thioparus* grows under near neutral conditions. *T. ferrooxidans* is also a strict aerobe and grows under acidic conditions. It oxidizes ferrous iron (Fe^{2+}). It is used for commercial leaching of ores and is responsible for acidity of water in mines. *T. denitrificans* oxidizes sulphur to sulphate utilizing nitrate as electron acceptor. Its activity results in loss of soil nitrogen as gas (denitrification).

$$2NO_3^- + S + H_2O + CaCO_3 \longrightarrow CaSO_4 + N_2$$

Sulfolobus and *Thermothrix* occur in hot sulphur springs, the former under acidic conditions and the latter at near neutral conditions. Both these, and *T. denitrificans* are capable of chemolithotrophic growth.

The sulphur-oxidizing bacteria can use different forms of sulphur (H_2S, sulphides, S, thiosulphate etc.) to form sulphate (SO_4)

1. **Strict aerobes:**
   - *T. thiooxidans* (electron donor) S
   - *T. thioparus* (electron donor) H_2S, sulphides, S°, $S_2O_3^{2-}$
   - *T. ferrooxidans* (electron donor) S, Sulphides, Fe^{2+}
2. **Facultative aerobes:**
   - *T. denitrificans* H_2S, S°, $S_2O_3^{2-}$
   - *Thermothrix* H_2S, $S_2O_3^{2-}$, SO_3^-
   - *Sulfolobus* H_2S, S°
3. **Heterotrophs:** *Beggiatoa* (filamentous, gliding bacterium) needs organic sources (acetate, succinate, glucose).

Hydrogen-oxidizing bacteria. Some Gram-positive and Gram-negative bacteria can obtain energy solely by utilizing H_2 as electron donor and O_2 as electron acceptor. They are **facultative lithotrophs**.

$$2H_2 + O_2 \longrightarrow 2H_2O$$

Pseudomonas flava and *Alcaligenes eutrophus* are examples of Gram-negative bacteria and *Bacillus schlegelii* and *Arthrobacter* sp. of Gram-positive bacteria. They possess membrane bound hydrogenases. Some of them (*P. carboxydovorans*) also oxidize CO to CO_2.

Iron-oxidizing bacteria. Some bacteria oxidize ferrous iron (Fe^{2+}) to ferric iron (Fe^{3+}). This process is important because it causes leaching of iron from the soil, especially in locations where iron ore is mined. The main organism responsible is *Thiobacillus ferrooxidans*. The ore pyrite is attacked forming ferrous iron and sulphuric acid. It is the acid that is responsible for leaching. The reactions are summarized below:

$$FeS_2 + 14Fe^{3+} + 8H_2O \longrightarrow 15Fe^{2+} + 2SO_4^{2-} + 16H^+$$

Gallionella under neutral conditions and *Sulfolobus* under acidic conditions also oxidize iron.

Magnetotactic bacteria. Magnetotactic bacteria (e.g. *Aquaspirillum*) show directed movement in a magnetic field, due to the presence of chains of a magnetic substance consisting of Ferric oxide (Fe_3O_4).

Carbon Assimilation in Autotrophs

Cyanobacteria, Rhodospirillaceae, Chromatiaceae, and *Chloroflexaceae* employ the Calvin cycle. The two key enzymes of the Calvin cycle, **phosphoribulokinase** and **ribulose-1, 5-biphosphate carboxylase** are present in these organisms. These enzymes are not present in *Chlorobiaceae*. Members of this family use the reductive TCA cycle for converting CO_2 to acetyl Co-A. This is possible because many reactions in the TCA cycle are reversible. These organisms utilize acetate, propionate or pyruvate.

Calvin cycle (overall reaction):

$$3CO_2 + 9ATP + 6NADH + 6H^+ \longrightarrow \underset{\text{(glyceraldehyde-3-phosphate)}}{GAP} + 9ADP + 8Pi + 6NAD +$$

The assimilation of CO_2 is by the Calvin cycle reactions, both in phototrophs and in chemolithotrophs. The pathway is illustrated in Fig. 41. It is carried out in **chloroplasts** in the eukaryotes. The nitrifying bacteria, thiobacilli and cyanobacteria possess inclusion bodies known as **carboxysomes** in which CO_2 fixation occurs.

In the Calvin cycle, ribulose-1, 5-biphosphate (RuBP) takes up CO_2 to form 3-phosphoglyceric acid (PGA) which is the primary product of CO_2 fixation. The enzyme involved is ribulose biphosphate carboxylase. Six ribose biphosphate molecules are rearranged in a series of steps involving C-3, C-5, C-6 and C-7 compounds to produce 6 molecules of 1-5-ribulose biphosphate. The enzyme catalyzing this reaction is phosphoribulokinase. Six molecules of ATP, formed in light reaction of photosynthesis are utilized in this process. One molecule of glucose is formed during the cycle. The cycle then repeats.

BIOLOGICAL NITROGEN FIXATION

Organic form of nitrogen is the major component of living organisms. It is the most abundant element in the atmosphere. However, it is not available to most organisms because of their inability to use it in the elemental form. Only a few groups of bacteria have the ability to combine it with other

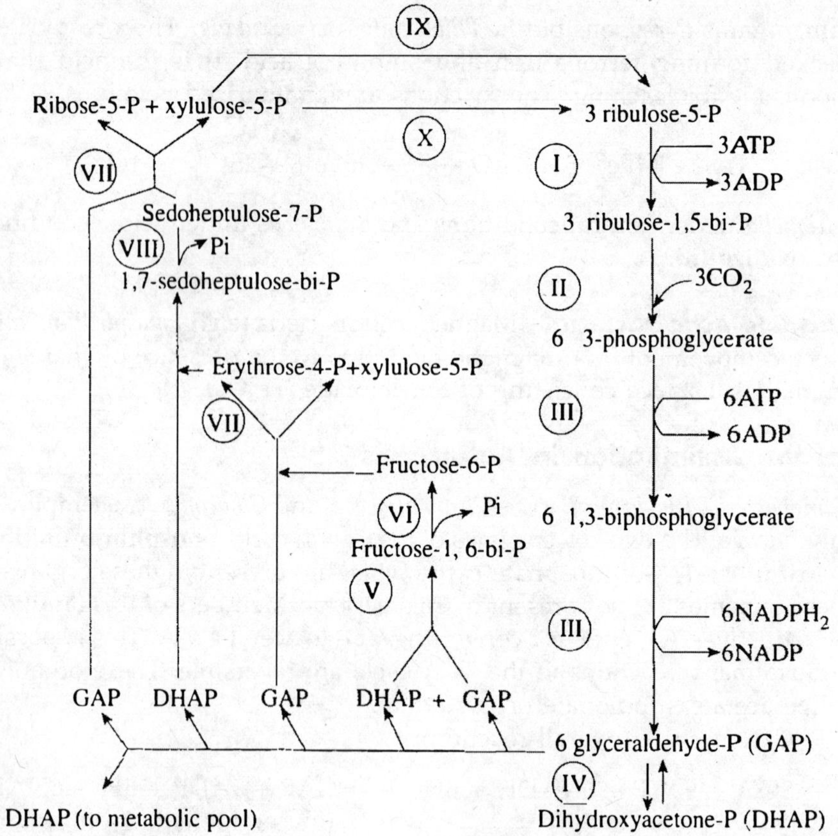

(I) Phosphoribokinase
(II) RuBP carboxylase
(III) Glyceraldehyde-3-phosphate dehydrogenase
(IV) Triosephosphate isomerase
(V) Aldolase
(VI) Fructose-1, 6-biphosphate phosphotase
(VII) Transketolase
(VIII) Sedoheptulase
(IX) Phosphopentoisomerase
(X) Phosphopentoepimerase

Fig. 41 Calvin cycle (reductive pentosephosphate cycle).

elements and use it directly. All other organisms depend on combined forms of nitrogen, mostly produced by these bacteria. Some of these bacteria are associated with plants especially legumes, and contribute to the enrichment of organic nitrogen. An experiment at Rothamsted Experimental Station (U.K.), extending over a century showed that in a wheat field which was regularly weeded, production was only 2 tonnes per hectare, while in a field left without weeding, it was 4 tonnes, indicating a gain of nitrogen by the wheat crop from the weeds which are mainly legumes.

A number of bacteria belonging to diverse groups have been shown to fix atmospheric nitrogen (Table 12). These organisms possess an enzyme system known as **nitrogenase complex** which reduces **molecular nitrogen** to **ammonia.** The process is energy-expensive, requiring about 20 ATP

Table 12 Nitrogen-Fixing Organisms (Partial List)

Organism	*Group*	*Comments*
A. Free living		
Azotobacter chroococcum	Gram-negative cocci	Aerobic, soil inhabiting
A. agills	-do-	-do-
Beijerinckia indica	Gram-negative rods	-do-
Derxia gummosa	-do-	-do-
Clostridium pasteurianum	Gram-positive rods	Obligate anaerobe, soil-inhabiting
Klebsiella pneumoniae	Gram-negative rods	Fixes nitrogen under anaerobic conditions
Anabaena spp. (and several other genera)	Cyanobacteria	Fixes nitrogen in heterocysts (*A. azollae* is symbiotic in *Azolla*)
Rhodospirillum	Purple photosynthetic bacterium	Anaerobic
Azospirillum spp.	Spirillaceae	Microaerophilic, fixes nitrogen in grass rhizosphere
B. Symbiotic		
Rhizobium leguminosarum	Gram-negative rods	Root nodules of pea
R. trifolii	-do-	Root nodules of clover
R. sesbaniae	-do-	Stem nodules of Sesbania
Rhizobium (cowpea group)	-do-	Nodulation in several tropical legumes
Bradyrhizobium japonicum	-do-	Soybean nodules
Frankia spp.	Actinomycete	Nodules in roots of non-leguminous plants

molecules to reduce one nitrogen molecule. Moreover, the enzyme system is sensitive to oxygen. Only those bacteria that have the enzyme system and also some mechanism to regulate oxygen in the cell have retained the capacity for nitrogen fixation. For instance, *Klebsiella pneumoniae*, a facultative anaerobe, fixes nitrogen when grown under anaerobic conditions but not under aerobic conditions. Cyanobacteria, which evolve oxygen in photosynthesis have heterocyst cells where nitrogen fixation occurs but no photosynthesis takes place. *Rhizobium* and *Frankia* fix nitrogen in symbiotic association with plants which have mechanisms for regulating oxygen supply. It is only some aerobic free-living nitrogen fixing forms such as *Azotobacter* that are able to separate oxygen and nitrogenase within the cell.

The nitrogenase complex essentially consists of two enzymes, the nitrogenase or **molybdoferredoxin** and the nitrogenase reductase or **azoferredoxin.** The two occur in roughly 1 : 2 proportion. Another enzyme, **hydrogenase,** is also found in many nitrogen-fixing organisms. The nitrogenases were first discovered in cell-free extracts of *Clostridium pasteurianum* by Carnahan and his associates in 1960. Since then, these enzymes have been isolated from most of the nitrogen-fixing organisms and they were found to resemble closely, hence the mechanism of fixation is considered to be essentially the same. The molecular weight of molybdoferredoxin is around 230,000 while that of azoferredoxin is around 70,000.

The nitrogen molecule is highly stable, the two atoms in it being connected by a triple bond. It requires very high activation energy for chemical reaction to occur. In the biological system, N_2 is reduced to **ammonia**, utilizing ATP and obtaining electrons from reduced ferredoxin or flavodoxin. Amino acids are produced by combining ammonia and organic acids.

The process also results in the formation of a hydrogen molecule, which is a wasteful product. However, the enzyme **hydrogenase** splits the molecule, recirculating the electrons and protons in the cell. In cyanobacteria, carbohydrates flow into **heterocysts** while amino acids produced in heterocysts flow back into vegetative cells. In the root nodules of legumes, host cells supply carbohydrates to *Rhizobium* and the amino acids are released into the host cell. The process is regulated by the switching off of nitrogenase activity under conditions of ATP starvation or ammonia accumulation.

The genetic mechanism of nitrogenase activity has been worked out in *Klebsiella pneumoniae*. A number of genes, known as **nif** genes are involved and they occur in more than one operon. Hence, this gene cluster is called a **regulon.** Its operation is controlled by another set of genes known as **ntr** genes. The **nif** gene cluster has been shown to consist of 17 genes in *K. pneumoniae*. They code for the functions such as formation of nitrogenase, synthesis of an iron-molybdenum co-factor, electron transfer, maturation of the enzymes and the regulation of expression of the concerned genes. Again, the **ntr** genes A, B and C regulate the expression of **nif** genes. The gene **gln A** controls the synthesis of glutamine synthetase, which is required for the assimilation of ammonia that is formed in nitrogen fixation. Figure 42 illustrates the pathway of nitrogen fixation, resulting in amino acid formation.

In root nodules, there is a close link between the host and the bacterium. Different degrees of specificity are known. Once the bacterium comes into contact with a compatible host cell, usually the root hair, it is drawn into an **infection thread** formed by the host. The presence of the bacterium triggers host cell division resulting in nodule formation. Simultaneously, the infection thread branches and the bacterial cells, which were multiplying, are drawn into host cells, surrounded by a membrane around them. The bacteria cease to multiply and are converted into irregular-shaped cells known as **bacteroids.** It is in this position that they start fixing nitrogen. Nodule cells show the presence of new proteins known as **nodulins,** one

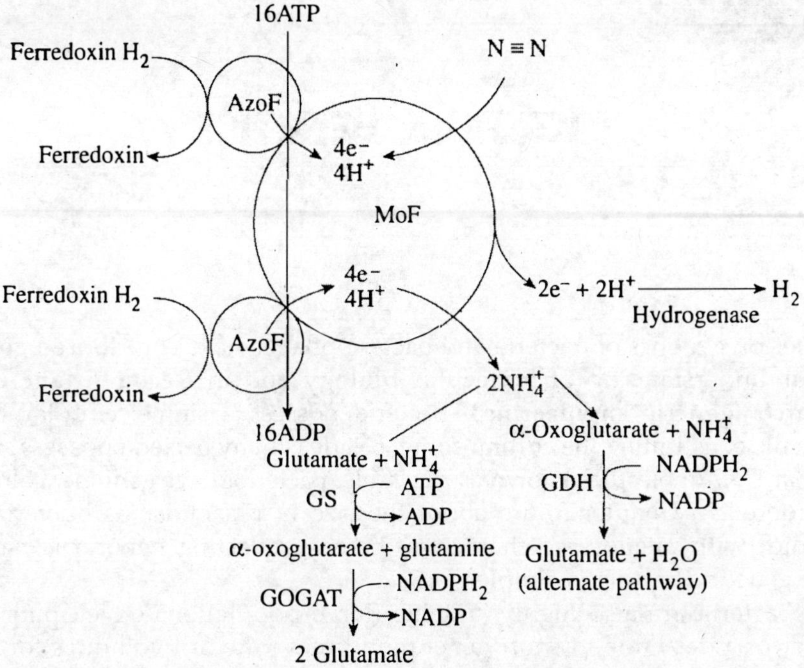

AzoF = Azoferredoxin (nitrogenase reductase)
MoF = Molybdoferredoxin (Nitrogenase)
G–S = Glutamine synthetase
GOGAT = Glutamine-emide-2-oxoglutarate aminotransferase
GDH = Glutamine dehydrogenase

Fig. 42 Nitrogen fixation and assimilation in legumes.

of which is **leghaemoglobin.** It regulates oxygen supply to the cell, facilitating nitrogenase activity. The presence of leghaemoglobin can be seen in nodule sections which appear pinkish red.

The root nodule bacteria of legumes are now grouped in two genera. *Rhizobium* is fast growing in culture while *Bradyrhizobium* is slow growing. The **nif** genes of the former are located in a large plasmid, while those in the latter are on the main chromosome. There are also **nod** genes in both, which specify nodulation by the host. The genes specifying leghaemoglobin seem to be present in both the host and the bacteria, hence nodulation occurs only after infection. Moreover, only some host cells, generally polyploid, take infection. The complex interaction between host and the bacterium is responsible for the specificity that is observed.

Nitrogen fixation also occurs in some non-leguminous plants, particularly forest trees (*Alnus, Hippophae, Eucalyptus* etc.). The nodules of these plants harbour an actiomycete, *Frankia*, which fixes nitrogen in the same way as *Rhizobium*.

7

MICROBIAL GENETICS

Studies on genetics of bacteria and bacteriophages have contributed greatly to our understanding of **molecular biology** and in **Recombinant DNA research (genetic engineering)**. Bacteria possess a single, circular DNA molecule constituting the **chromosome**. Besides, many bacteria possess small, circular DNA molecules known as **plasmids**. Bacteriophage genomes (already considered in Chapter 3) are about the size of plasmids. As bacteria are haploid, with phenotypic characters directly reflecting genotypic nature, their genetic analysis is simpler.

Bacteria are also highly suitable (large populations developing in a matter of a day) due to short generation times (one or two hours for each doubling of population). They can be cultured on defined media in the laboratory and their nutritional requirements or response to drugs and environmental changes can be precisely identified. Mutations naturally occurring or induced by physical or chemical agents **(mutagens)** provide a wide range of variation in bacteria, facilitating genetic analysis. Mutant strains of bacteria, bacteriophages and plasmids are important in genetic studies, hence they are briefly discussed below, prior to a consideration of genetic mechanisms in prokaryotes.

MUTATIONS

A **mutation** is recognized as a **sudden, heritable change** in the phenotype of an organism. Mutations naturally occur in populations in low frequency, generally one in 10^6 to 10^9 cells. For experimental purposes mutants are produced by subjecting the populations to certain physical or chemical treatments. As a result of such treatments, mutants may occur at frequencies of one in 10^3 to 10^6. In bacteria, phenotypic characters such as change in nutritional requirements, temperature optima, response to drugs, colony characters, host range etc. are brought about by mutations. The bacterial strains isolated from nature are known as **wild strains**, while the others are referred to as **mutants**. Bacteria which grow on a carbon source and minerals are referred to as **prototrophs**. Those that require additional nutrients for their specific requirements are referred to as **auxotrophs**, and are designated with suitable symbols. For instance, a strain of *E. coli* requiring the amino acid **leucine** in the medium is designated as **leu**$^-$, in contrast to the prototroph which does not require leucine, and is designated as **leu**$^+$. Several types of auxotrophs have been isolated and extensively used in the study of microbial genetics and molecular biology.

Since the frequency of occurrence of mutants is very low, suitable detection techniques are employed. A commonly employed technique is the **replica plating technique**. For instance, a strain exposed to a mutagen is plated out on a **complete medium** which contains a range of supplement nutrients, besides carbon source and minerals. Plates containing the **minimal medium**, containing only carbon source and minerals, and plates containing different supplements to the minimal medium are also prepared. An isolate exposed to a **mutagenic agent** is plated out on a complete medium so that prototrophs as well as auxotrophs develop. A velvet cloth attached to a wooden disc of a size that fits into a Petri dish is taken and sterilized. It is pressed on the surface of the complete medium plate on which bacterial colonies had developed. The disc is then pressed successively on plates containing the minimal medium and other nutrient supplements, as shown in Fig. 43. The colonies developing on the minimal medium are prototrophs, and those developing on plates with nutrient supplements are the auxotrophs

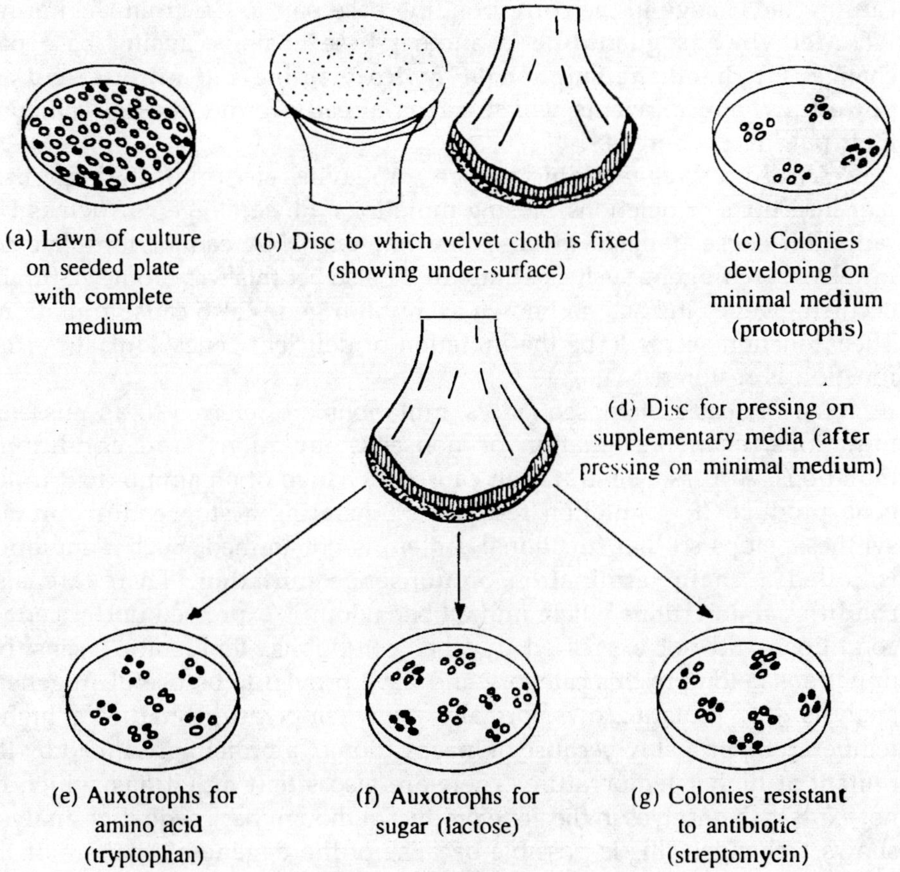

Fig. 43 Replica plating technique.

for those particular nutrients. Auxotrophic mutants are thus detected and subsequently isolated on suitable media.

Different mutagens (agents that cause mutations) affect the bacteria in different ways. A change in base sequence in DNA gives rise to a mutant phenotype. If the change is only in one base pair it is called **point mutation**. It may be due to a base **insertion**, base **deletion** or base **substitution**, as illustrated below. Base substitution (replacement of one base by another) is a more frequently occurring change.

DNA sequences

- Wild type — ATGA, CCAG, GTC
- Base substitution — ATGA, CTAG, GTC (T replacing C)
- Base addition — ATGA, CACA, GGTC (addition of A)
- Base deletion — ATGA, CAG, GTC (deletion of C)

A commonly employed mutagen is **5-bromouracil**. It is an analog of the base thymine (an analog is a substance which closely resembles the normal one in structure but is functionally different). It can replace thymine in the DNA molecule. 5-bromouracil bonds with cytosine (instead of adenine) causing the change in the corresponding base pair to GC from the normal AT. **Methylnitrosoguanidine** is another base analog, causing base pair change. **Ethylmethane sulphonate**, **hydroxylamine** and **nitrous acid** are powerful chemical agents which act on already formed bases, changing their pairing properties.

Acridine dyes (acridine orange, proflavins etc.) induce single base pair insertions or deletions causing mutations. Physical agents such as **UV radiation** cause damage to bases so that the DNA cannot template for mRNA. **Carcinogens** such as **aflatoxin B1** also act this way. Some naturally occurring genes in bacteria known as **mutator genes** also cause mutations. Their function seems to be the mutation of deficient genes so that normal function is restored.

On the basis of consequences, mutations are referred to as missense mutations, chain termination or nonsense mutations, and conditional mutations. A **missense mutation** causes a change of an amino acid in the gene product. If a mutation results in generating a **stop codon**, protein synthesis stops so that functional protein is not formed. Such a mutation is called a chain termination or **nonsense mutation.** There are also **conditional mutations** where mutant behaviour is expressed under certain conditions and not expressed in other conditions. Temperature-sensitive mutations belong to this category and have proved to be useful in genetic analysis. A TS mutant grows normally at low temperature but dies at higher temperature, probably because of inactivation of a protein produced by the mutant at higher temperature. There are also **silent mutations** which do not express themselves in the gene product, although **base sequence analysis** shows a change. This is possible because of the **degenerate** nature of the genetic code. In the **triplet code** the third base is not as essential as the first two and a change in the third base may not change the amino acid for which it codes.

Sometimes, **back mutation (reversion)** occurs resulting in the development of a wild type from a mutant. This may be due to a suppressor gene at some other site in the chromosome reverting the mutant gene function.

GENES AND GENETIC CODE

Before the molecular structure of DNA was known, a **gene** was defined as an entity that **confers a single trait** on an organism, which can undergo recombination. Techniques of complete sequencing of nucleotides in DNA and RNA are now available and the triplet code has been established. As the DNA functions by triplets of nucleotides coding for protein synthesis, each triplet coding for a single amino acid in a polypeptide chain, a gene is now defined as a **linear sequence of nucleotides with a starting point and an end point**. Such a segment of DNA is also called a **cistron**. Besides proteins, DNA also codes for rRNA and tRNA. Some proteins have more than one polypeptide chain, hence more than one cistron is required to code for such proteins.

There are 20 different amino acids in proteins. Each is coded by a set of three nucleotides occurring in a sequence (because sequences of two nucleotides can only code for 16, as there are only four different nucleotides in DNA). Different combinations of the four nucleotides in triplets can code for 64 combinations (4^3). This is more than what is required to code for the 20 amino acids. It has been found that each amino acid has two to four triplets coding for it. Therefore, the genetic code is said to be a **degenerate code**. In triplets coding for each amino acid, the first two nucleotides are always common. For instance, the triplet codes for alanine are GCU, GCC, GCA and GCG. Besides the codons for the amino acids, there are three triplets that code for termination of sequences. They are called **stop** or **nonsense codons**. Table 13 illustrates the genetic code.

The mechanism of protein synthesis involves **transcription** of codons on to **messenger RNA (mRNA)**. Ribosomes are present in the cell and the mRNA formed is arranged on a set of ribosomes. There are several species of **transfer RNA (tRNA)** which bring the amino acids to form a polypeptide chain on the basis of codes present on the mRNA. **Ribosomes** are made up of two subunits of ribosomal RNA (rRNA) and proteins. The ribosomes of prokaryotes are of the 70 S type and consist of two nucleotide subunits, a protein and also enzymes needed to form peptide bonds between amino acids.

Transcription

The formation of mRNA molecule on a DNA template is known as transcription. The first step in the formation of mRNA is the attachment of the RNA polymerase to a region in DNA, which is known as **promoter**

Table 13 The Genetic Code

First position (5′ end)	*Second position* U	C	A	G	*Third position (3′ end)*
U	Phe	Ser	Tyr	Cys	U
	Phe	Ser	Tyr	Cys	C
	Leu	Ser	<u>Stop</u>	<u>Stop</u>	A
	Leu	Ser	<u>Stop</u>	Try	G
C	Leu	Pro	His	Arg	U
	Leu	Pro	His	Arg	C
	Leu	Pro	Gln	Arg	A
	Leu	Pro	Gln	Arg	G
A	Ile	Thr	Asn	Ser	U
	Ile	Thr	Asn	Ser	C
	Ile	Thr	Lys	Arg	A
	[Met]	Thr	Lys	Arg	G
G	Val	Ala	Asp	Gly	U
	Val	Ala	Asp	Gly	C
	Val	Ala	Glu	Gly	A
	[Val]	Ala	Glu	Gly	G

Those shown in boxes are initiation codons and those underlined are stop (termination) codons.

region. This event opens up the two strands of DNA at the region, forming a loop. The RNA polymerase initiates the formation of the nucleotide sequence. It moves along the DNA nucleotides in the loop, adding complementary nucleotides in a sequence. As the RNA polymerase moves, the two strands of DNA close. On reaching a termination codon, further extension of mRNA stops and it is released. The mRNA formation is illustrated in Fig. 44.

Translation

The synthesis of a polypeptide is known as translation. It involves three steps, viz. initiation, elongation and termination. The first step is the binding of mRNA to a set of ribosomes, which are collectively known as **polyribosomes** or **polysomes**. The amino acids get attached to tRNA molecules. The tRNA molecules are folded in such a way as to form a clover-leaf structure. This is brought about by the formation of complementary nucleotides opposite to each other, to form a double strand, while loops are formed where complementarity is absent. The nucleotide triplet in the middle loop is known as an **anticodon**. It recognizes the triplet code on mRNA. The amino acid is attached to the open 3' end of the tRNA molecule which has the C–C–A sequence.

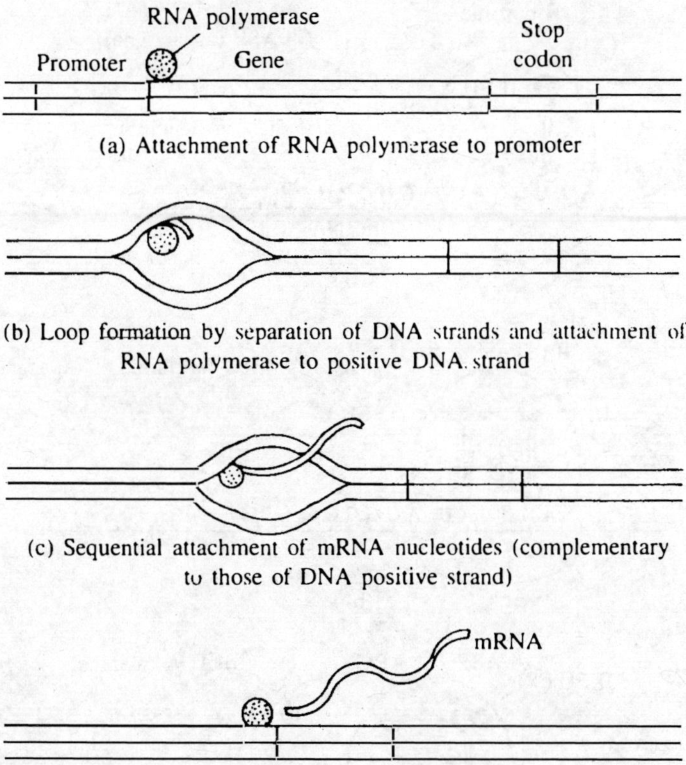

(a) Attachment of RNA polymerase to promoter

(b) Loop formation by separation of DNA strands and attachment of RNA polymerase to positive DNA strand

(c) Sequential attachment of mRNA nucleotides (complementary to those of DNA positive strand)

(d) Release of mRNA from RNA polymerase

Fig. 44 Formation of mRNA on DNA template (transcription).

The elongation of peptide involves attachment of amino acids by the formation of peptide bonds between them. This is brought about by the moving of the mRNA molecule along the ribosomes, bringing successive codes into position. In the termination stage, the polypeptide that is formed gets separated from the ribosome. Figure 45 illustrates the translation process.

Plasmids

Plasmids are small DNA molecules occurring in many bacteria and a few eukaryotes. They are circular and supercoiled and usually separate from the main chromosome. Some plasmids occur only as one or two copies in a cell, while others occur in a large number of copies. Some plasmids can get integrated with the main chromosome or exist independently. Such plasmids are called **episomes** (e.g. F-Factor in *E. coli*). Because of their small size, they can be isolated intact from the cell. They carry a small number of genes which impart useful properties to the cell. The plasmids replicate independently at the time of cell division, so that the daughter cells also have them. Although they are useful to the cell, they are not strictly essential.

On the basis of functions of plasmids (the genes present in them), they are grouped into **conjugative** plasmids, **drug resistance** plasmids

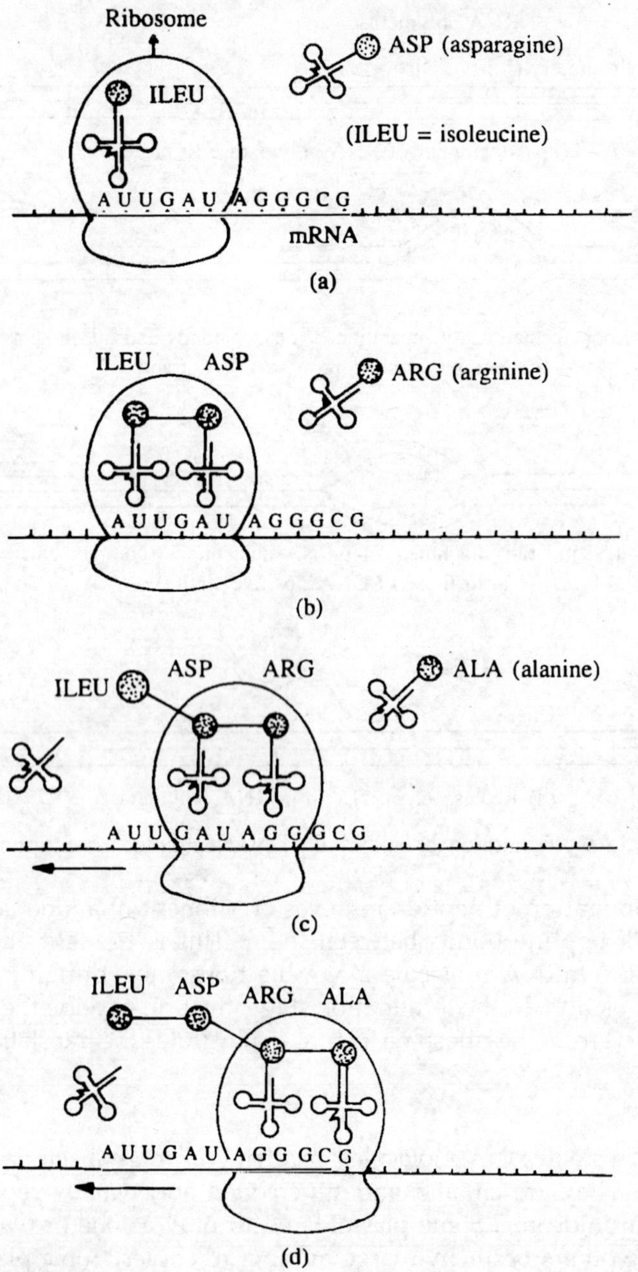

Fig. 45 Translation process (formation of polypeptide chain with mRNA attached to ribosome).

(R-plasmids), **metabolic** plasmids and **virulence** plasmids. The properties of some important plasmids are listed in Table 14. Bacterial cells can be made to lose the plasmids by a process of **curing**. A comparison of properties of plasmid containing and cured cells reveals the genes carried by the plasmids.

Table 14 Properties of Some Important Plasmids of Bacteria

Plasmid type	*Designation*	*Usual copy No.*	*Hosts*	*Phenotypic features*
F-factor	F^+	1 to 3	*E. coli, Salmonella*	Sex pilus formation, conjugation
R-plasmids	RP_4	1 to 3	*Pseudomonas and other Gram '–' forms*	Resistance to antibiotics (kanamycin, ampicillin etc.)
	PSH_6	1 to 3	*Staphylococcus aureus*	Resistance to antibiotics (kanamycin, gentamycin etc.)
Col-plasmids	Col E_1	10 to 30	*E. coli*	Production of colicin E_1
	Col E_2	10 to 15	*Shigella dysenteriae*	Production of colicin E_2
Metabolic plasmids	SAL		*Pseudomonas spp.*	Degradation of salicylic acid
Virulence plasmids	P 307		*E. coli*	Production of enterotoxin

The R-plasmids carry genes for resistance to one or more antibiotics and sulphonamides. There are **col plasmids** which carry genes for **colicins**. A col plasmid-containing cell prevents the growth of related bacteria, if they do not possess such col plasmids. The crown gall bacterium *(Agrobacterium tumifaciens)* carries a **ti-plasmid**, which triggers tumour formation in host plants. The ti-plasmid is an episome and gets integrated with the host chromosome. This property of integration with eukaryotic chromosomes has opened up the possibility of using this plasmid for transfer of genes in higher plants.

In *Escherichia coli*, a fertility factor, known as **F-factor** was the earliest plasmid discovered in bacteria. During conjugation (which will be described later), the cell containing it (F-factor) behaves as a **donor** (male) and transfers a copy of it to a **recipient** (female) cell. The F-factor replicates by a mechanism known as a **looped rolling circle** mechanism, (as illustrated in Fig. 46), one copy moving into the recepient and one being retained in the donor cell. **Thus, the recipient cell is converted into a donor cell.** Sometimes, the F-factor gets integrated with host chromosome and is carried from generation to generation. Such a cell is known as an **HFr strain**. When the F-factor separates from a host chromosome, it carries a few host genes located adjacent to it on the main chromosome. Hence, the F-factor plasmid has also been used in gene transfers.

Transposons

Transposable elements in bacteria were detected as polar mutations in **gal**

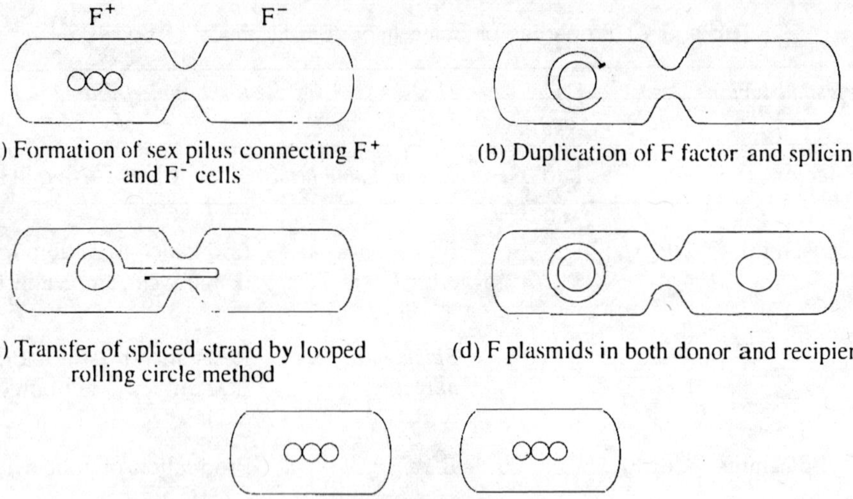

(a) Formation of sex pilus connecting F^+ and F^- cells

(b) Duplication of F factor and splicing

(c) Transfer of spliced strand by looped rolling circle method

(d) F plasmids in both donor and recipient

(e) Supercoiling of plasmid DNA

Fig. 46 Replication of F plasmid.

operon of *E. coli*. Analysis of DNA sequences in the mutants showed that such mutants had a segment of DNA inserted into the gene. Such segments were called **insertion elements** (IS elements). These were shown to have a class of sequences and they were called transposable elements or **transposons**. They were shown to range in length from a few hundred to a few thousand nucleotide pairs. A common feature of all IS elements is the presence of **inverted repeat sequences** at terminals, as shown below:

A G T C	G A C T
T C A G	C T G A

The main body of the transposon contains either transcription termination signals or stop codons. In *E. coli*, several transposons were found in the main chromosome and also in the F plasmid. They are able to move from position to position, hence the term transposon.

Transposons were also found to carry additional genes, especially antibiotic resistance genes. Those containing the antibiotic resistance gene flanked by identical IS elements on either side, are said to belong to a **composite family.** They can insert themselves into other genes, in either **direct** or **inverted** manner because their terminals have inverted repeats, as explained above. There are larger transposons which are grouped under the TN3 family. They carry genes for β-lactamase (which confers resistance to ampicillin) flanked by the IS elements.

Genetic analysis of transposon-containing cultures showed that mutants caused by them have two copies of the transposon, one at the original site and another at a new site. Sometimes more than two copies are present. At the time of insertion, a sequence in the chromosome duplicates. Such a sequence is called a **target site**. The copy of the transposon gets inserted between the duplicate sequences of the target site. Hence, the transposon is retained at the original site and it is the copy which gets inserted at a different site, leading to mutations.

Transposons were employed to bring about certain genetic phenomena such as inversions, deletions and fusion of DNA molecules. Formation of HFr cells is a result of the transposon in the F-plasmid which facilitates its integration with the main chromosome. **Gene amplification** occurs when multiple copies of a transposon (containing the drug resistance gene) are formed in the cell. This phenomenon causes increased drug resistance of the bacterium. Figure 47 illustrates the steps in the insertion of a transposon on the main chromosome.

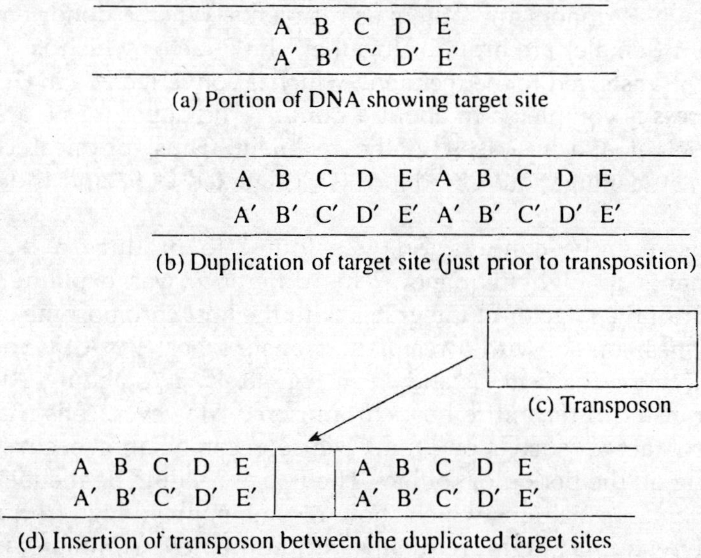

Fig. 47 Insertion of a transposon between duplicated target sites.

RECOMBINATIONS IN BACTERIA AND PHAGES

Transformation

The first indication that a chemical substance is genetic material came from the study of Griffith (1928) on pneumococci *(Streptococcus pneumoniae)*. A non-pathogenic strain produces rough colonies while a pathogenic strain produces smooth colonies. This is due to the absence of the capsule in the former and its presence in the latter. When heat-killed cells of the smooth colony were mixed with those of the rough colony and plated out, some colonies that developed were rough while some were smooth. The latter proved to be pathogenic. Some substance from heat-killed cells had transformed the rough colony-producing cells to smooth colony-producing cells. This phenomenon was called **transformation**. Later it was shown that the transforming principle is a nucleic acid (Avery et al., 1944). Proof that DNA is the genetic material was obtained by Hershey and Chase (1952) who showed that radioactively labelled DNA of phage T2 entered *E. coli* cells leaving behind the protein coat outside the cell. In more recent times, the transformation principle has been employed for the transfer of genes. It was found that precipitation of DNA molecules with calcium phosphate facilitates their uptake by cells.

Conjugation in Bacteria

Lederberg and Tatum (1946) observed that when strains of *E. coli* auxotrophic to different nutrients were mixed and grown for several generations, some prototrophs were produced. They assumed it to be due to genetic recombination and termed it as **conjugation**. Later it was shown that some cells behave as donors and others as recipients. When a donor (male) and a recipient (female) are brought together, the F-factor which is a plasmid is quickly transferred to the recipient, which is converted to a **male** (donor). This process is completed in about a minute. The donor forms a **sex pilus** with which it attaches itself to the recipient. Thus, physical contact is necessary in conjugation. The donor is designated as F^+ and the recipient as F^-.

Subsequently, conjugation was found to produce a number of recombinants in high frequency. This difference was explained by the behaviour of the F-factor. It integrates with the host chromosome and when it establishes contact with a recipient, transfers host DNA by the looped rolling circle mechanism, as shown in Fig. 48. It takes about 100 minutes for the transfer of the entire host chromosome. However, cells may not be in close contact for such a long time. Hence, recombination occurs in genes depending on the period of contact. The F-factor would be the last to enter the recipient. Such strains which show recombinants in high frequency are called **HFr strains (high frequency recombinants)**. Wollman and Jacob (1952) developed a method of interrupting the conjugation process at various time intervals after mixing the donor and recipient cultures. By analyzing such cells they were able to map the genes in the *E. coli* chromosome. The earliest transferred genes would be at one end of the chromosome while those transferred later would be close to the F-factor. The F-factor is the last to be transferred, only when the conjugation process is complete. (Such cells again behave as HFr strains). Positions of some important genes mapped on *E. coli* chromosome by this technique are illustrated in Fig. 48.

Transduction

Bacterial genes are sometimes transferred through bacteriophages. This is known as **transduction**. When phage nucleic acid is synthesized in the host cell, occasionally a small fraction of host DNA gets incorporated in the phage DNA due to error in virus replication. An equal amount of phage DNA is left behind because the capsid of the phage cannot hold a larger quantity of nucleic acid. Such phages are known as **defective phages**. When

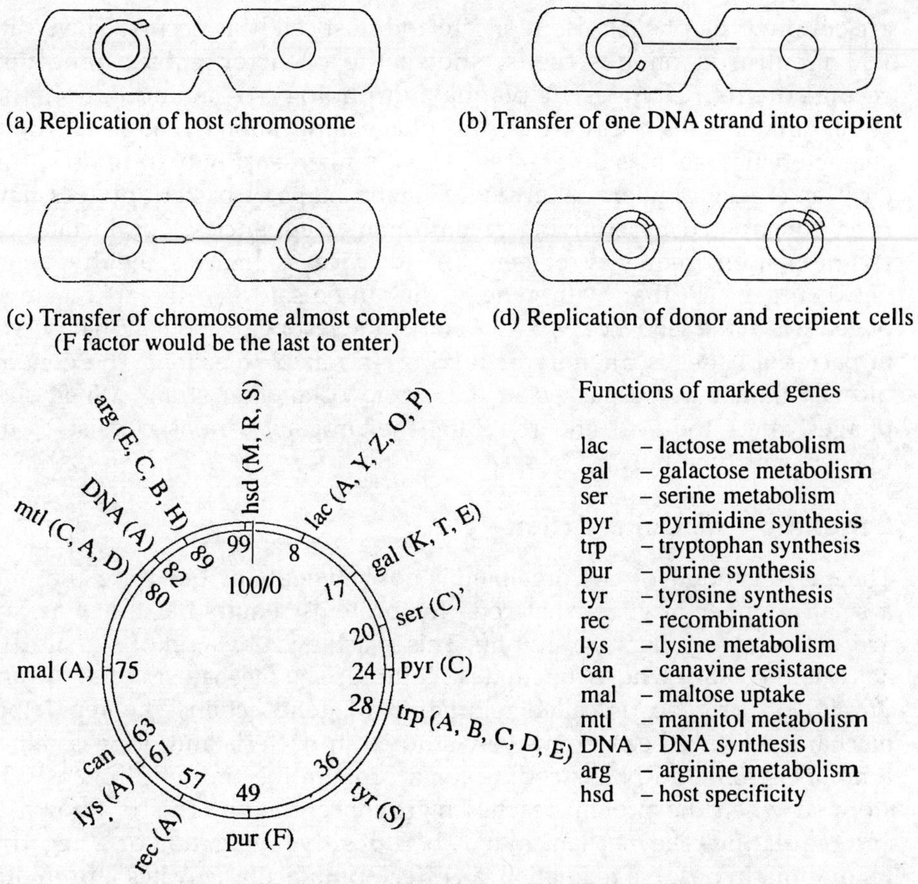

(a) Replication of host chromosome (b) Transfer of one DNA strand into recipient

(c) Transfer of chromosome almost complete (F factor would be the last to enter) (d) Replication of donor and recipient cells

(e) Partial genetic map of *E. coli* showing the position of some genes

Fig. 48 Conjugation with HFr strain and a partial genetic map of *E. coli.*

such a phage infects a host, the DNA fraction of the host recombines with the host chromosome, resulting in gene transfer. This phenomenon is called **transduction** and the phage is known as a **transducing phage**.

If there is no specificity and if any part of the host chromosome is transduced, the transduction is said to be **generalized**. If, on the other hand, the transducing phage carries only a particular part of the host chromosome, the phenomenon is called **specialized transduction**. This occurs in lysogenic phages (e.g. phage λ) because when they are in prophage state they get integrated with the host chromosome at a specific site. When lytic state occurs they may carry an adjacent host chromosome fraction. For instance, the phage lambda genome gets integrated next to **gal** and **bio** genes. When the phage leaves the cell it carries those genes.

Recombination in Bacteriophages

When more than one virion infects a host cell, recombination between the phages can occur. When bacterial cells containing such mixed infections are cultured the phages also replicate and host cell lysis occurs. The lysate, when cultured on host cells, shows the occurrence of **phenotypic recombinants**. Changes in plaque morphology is an indication that recombination has occurred during phage replication. Changes in other characteristics, such as differences in host range, sensitivity to temperature etc., have been similarly analyzed. Genetic maps of bacteriophages have been constructed by studying **recombinant frequencies.** For instance, if recombination frequency of gene 'A' is twice as high as the frequency between gene 'B' than with gene 'C', it can be said that the arrangement of genes is in the sequence A, B, C. Another interesting phenomenon observed in bacteriophages is **phenotypic mixing**. In mixed infections, the genome of one strain may be packed in the capsid of another strain. When such phages infect the host and replicate, the progeny virions consist of the nucleic acid type only.

Regulation of Gene Action

The entire genome of any organism is not active at any one time and only a small number of genes that code for proteins required at that time are active while the others are shut off. This is achieved by a set of signals that activate the genes whose function is required. The mechanism of switching on and off of genes is called **regulation of gene action**. The regulatory mechanisms have been extensively studied in bacteria and phages. When a monocistronic mRNA is coding for a protein, the transcription may be stopped when the protein reaches high concentration. This is known as **autoregulation.** The mechanism may be a **positive regulation** or a **negative regulation**. In positive regulation, an **effector molecule** activates a **promoter gene** that induces transcription. In negative regulation, an **inhibitor is bound to the gene preventing transcription.** When an **inducer substance** is present, the inhibitor is removed from the gene and transcription occurs.

One of the first regulatory mechanisms discovered is **lac-operon** in *E. coli*. It is a system having both positive and negative regulation functions. The enzymes required for lactose metabolism by this bacterium are **lactose permease**, which moves lactose into the cell across the membrane, and **β-galactosidase**, which breaks down the lactose molecule. The operon has three structural genes, "z, y, and a", one **repressor**, one **promoter** and one **operator** gene. The 'z' and 'y' code for β-galactosidase, and 'a' for another enzyme. Gene "i" codes for a repressor which binds to gene "O", the operator gene. This is negative regulation.

When *E. coli* is growing on glucose, cyclic AMP and a CAP protein (catabolic activator protein) are present in low concentrations. When glucose is exhausted and lactose is present, the two compounds increase in concentration and bind to the promoter gene causing it to transcribe. This is positive regulation. Figure 49 illustrates the two aspects of regulation.

A different type of operon functions in **biosynthesis of tryptophan** in *E. coli*. In this system, the product itself acts as a repressor. When tryptophan is present in the growth medium in adequate amount its biosynthesis is

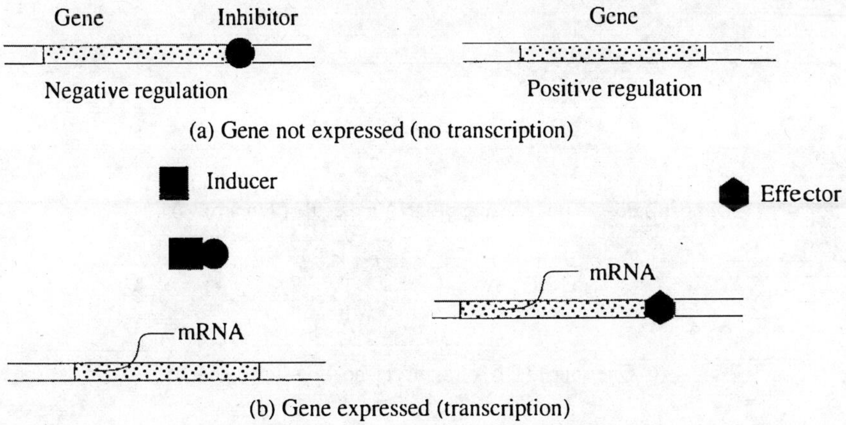

Fig. 49 Lac operon in *E. coli*.

inhibited. A substance known as **trp-aporepressor** is present in the cell. When tryptophan is present, the aporepressor binds to it to produce an active repressor, which binds to the operator gene so that transcription is prevented. When tryptophan is absent, transcription occurs because the active repressor is not formed. This system is known as **on-off** regulation.

RECOMBINANT DNA

The discovery of **restriction endonucleases** (generally referred to as restriction enzymes) in bacteria laid the foundation for recombinant DNA technology. Restriction enzymes split DNA molecules at specific sites. There are another set of enzymes known as **DNA ligases**, that bind DNA fragments. With those enzymes it has been possible to connect DNA fractions from different sources. A few hundred restriction enzymes have been isolated from bacteria. Their natural role is to protect the bacterium from foreign DNA (such as that of viruses). The bacterial DNA itself is protected by some modification.

The technique of breaking up of the DNA molecule is called **splicing**. Different restriction enzymes split the molecule between specific nucleotides. The double helical structure of the DNA fragment is retained, except at its **split ends** where a few nucleotides are exposed. These ends are known as **sticky ends**. DNA ligases can join the sticky ends of DNA fragments if the nucleotides are complementary, as illustrated in Fig. 50.

The DNA fragments from different sources can thus be joined. They are then introduced into suitable bacterial cells. For enabling the entry of foreign DNA, the desired fragments are joined with plasmid or phage DNA, which can enter the host cell. Such vehicles that introduce foreign DNA into cells are known as **cloning vehicles** or **vectors.** The vectors together with foreign DNA can reproduce in the subsequent cell generations. Such cells carrying the foreign DNA are known as **clones** and the process is referred to as **gene cloning**.

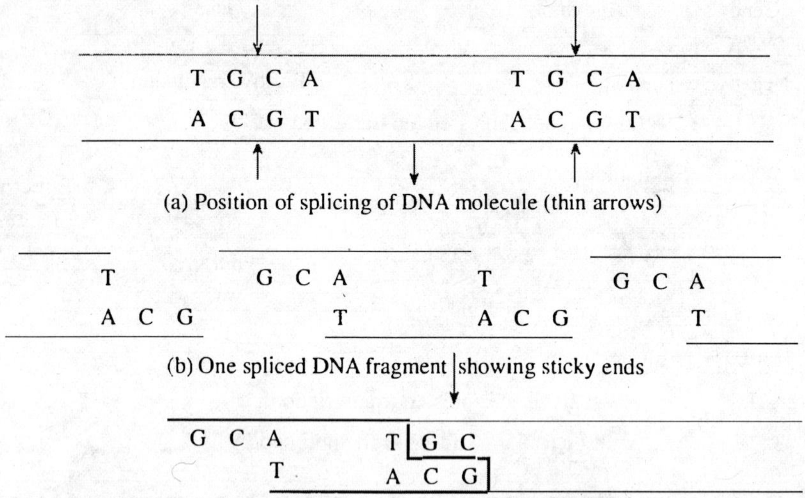

(c) Position of ligation of two DNA fragments (one of them shown in thick line for recognition)

Fig. 50 Splicing of DNA molecule by a restriction enzyme and ligation of two fractions.

Nowadays it is possible to isolate minute quantities of proteins and determine their amino acid sequences. This information is used to construct **probes** that can identify the genes which code for those proteins. Because the genetic code is known, nucleotide sequences that code for the amino acids of those proteins can be isolated from a source of genetic material. For instance, **human DNA** is spliced into fragments and maintained under appropriate conditions. It represents what is called a **gene library**. They are actually bacterial cells, each having one human DNA fragment. The fragments are introduced into the cell through the vectors. The vectors are either plasmids or modified phages such as a **phage lambda cosmid** or a **mini Mu phages** (already explained in the chapter on viruses). Once a gene is introduced into a bacterium, the cloned bacterium is multiplied so that a large quantity of the product can be collected.

The next step, i.e. identifying the gene in the cloned gene library is the more difficult part. Radioactively labelled DNA probes are employed. Under appropriate conditions, the probe binds with the gene. The particular clone containing the gene is thus recognized. It is also possible to analyze protein products in cloned cells and identify the clone which forms the gene product for the genes that code for known proteins. A third method used is the **immunological approach** where the product is subjected to **antibody assay**. The clone that contains the **antigen** is thus identified.

To be functional, the clone must contain the entire operon, with promoter, activator and structural genes, as well as start and stop signals. Only those clones that contain the entire set would be identified. Further propagation of such a clone is easy, enabling the collection of the protein product in large quantities. Some of the clinically useful proteins such as **insulin** (for controlling diabetes), the **human growth hormone** (for correcting dwarfism etc.) and **interferons** (for controlling certain cancers and virus diseases) are now commercially available and are in use. They are all produced in cultures of recombinant *E. coli* clones. Besides *E. coli*, yeasts and fungi can also be used as cloning vehicles. For certain complex proteins, which cannot be synthesized by *E. coli*, **mammalian cell lines** grown in tissue culture are used as cloning vehicles. The **t-DNA** of crown gall pathogen (*Agrobacterium tumefaciens*) has been useful as a vector in transfer of genes into plant cells. For instance, plant cells were cloned with a gene imparting resistance to insect pests, and the plants were raised from such cloned cells, giving rise to pest-resistant plants.

8

IMMUNOLOGY

Humans and higher animals possess several mechanisms for protection against invading microorganisms. Skin is the first barrier. Lactic acid present in sweat and sebaceous secretions is inhibitory to them. Tears, saliva and urine wash them down. Epithelical cells of the mucous membranes prevent adherence of microbes to tissues. Antagonistic microbes in the intestine also exert inhibitory action. In spite of these defences, several microbes invade the body and the immunity mechanisms then come into play to protect the organism.

The study of physiological processes that generate specific responses to infection or entry of foreign substances into the body is called **immunology**. The leucocytes of blood (white blood cells) are involved in producing immunity. A microbial cell or foreign substance that causes an immune response is known as an **antigen**. The response is of two types: In **phagocytosis** the microbial cell or foreign particle is engulfed and digested by a type of white blood corpuscle known as a phagocyte. In the second type, molecules that bind to the antigens are formed and inactivate the antigen. This is known as **natural immunity**. Figure 51 shows attachment of antibodies to an antigen.

The substances which elicit immune response are known as **immunogens**. They are usually proteinaceous or polysaccharide macromolecules. Some smaller molecules also trigger immune responses when they are linked to larger molecules in the body. Such substances are known as **haptens**. Many drugs, including penicillin act this way. The immunity brought about by phagocytosis is known as **cellular immunity** and that brought about by antibodies as **humoral immunity**. However, there is interaction between the cells that bring about these responses and the terms refer only to the end reactions.

CELLS OF THE IMMUNE SYSTEM

The various cell types involved in immunity arise from bone marrow stem cells. The phagocytes are of two types, **myeloid cells** and **mononuclear phagocytes**. The myeloid cells possessing prominent granules and referred to as **granulocytes**. **Neutrophils, basophils** and **eosinophils** belong to this category. The neutrophils are also known as PMN cells (Polymorpho nuclear

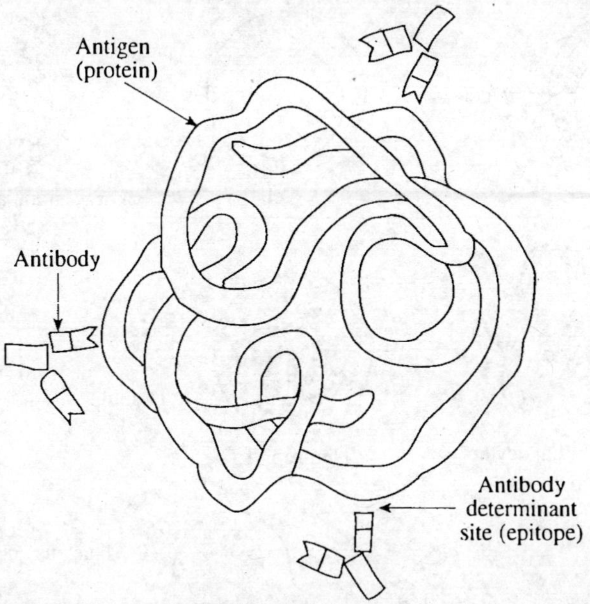

Fig. 51 Attachment of antibodies to an antigen (protein).

neutrophils) and generally as **polymorphs**. They are short-lived but constitute the major leucocyte population in the blood, due to their rapid turnover. The mononuclear monocytes which are long-lived and associated with tissues are known as **macrophages**. There is also a group of cells known as **large granular lymphocytes (LGHs)**, or **natural killer cells (NK cells)**. They can recognize tumour cells or virus-infected cells and destroy them.

Lymphocytes are responsible for initiating immune responses. They are distinguished into B-cells and T-cells. Both arise from bone marrow stem cells, but while B cells mature before they are released from bone marrow, T-cells mature in the thymus. The lymphocytes are spherical, with a large nucleus and a rim of cytoplasm. They are found largely in the liver, spleen, tonsils, adenoids and in lymphoid tissue. Figure 52 illustrates various cell types associated with the immune system.

There are certain soluble components which are also involved in the immune system and a group of about 20 serum proteins (CRP proteins) increase enormously following infection. They form the **complement system**. They control inflammations. They form a coating around microbes, a process known as **opsonization,** which helps in their recognition by the immune system. They attract phagocytes to the site of infection. They also damage the plasma membrane of Gram-negative bacteria causing cell lysis and also damage the viral envelopes. Platelets in blood whose normal function is the clotting of blood, also release complement proteins.

Cytokines are a group of glycopeptides involved in signalling between cells during immune response. (Cytokines produced by lymphocytes are known as **lymphokines.**) **Interferons** are proteins formed in virus-infected

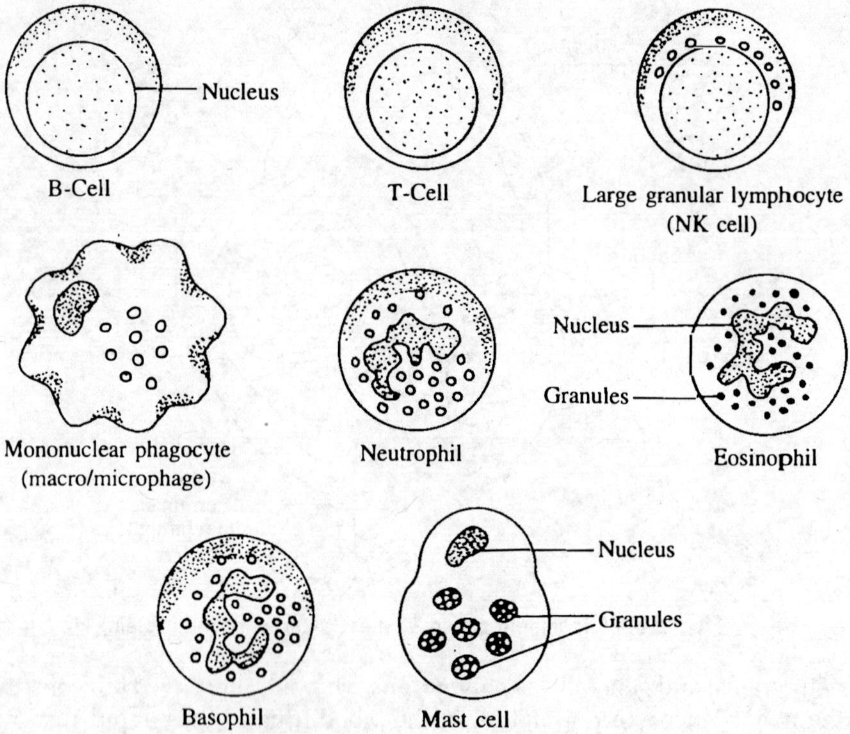

Fig. 52 Cell types participating in natural immune systems.

cells. They prove to be useful as antiviral and antitumour agents. **Interleukines** produced by T-cells are involved in the maturation of various cell types produced in bone marrow stem cells. Colony-stimulating factors (CSFs) stimulate the division of bone marrow stem cells. All these forms, cellular or molecular, constitute the immune system.

ANTIGEN RECOGNITION AND ANTIBODY FORMATION

The B-cells have cell surface immunoglobulins (gamma globulins) which have the capacity to bind antigens. B-cells are activated in the presence of the antigen and produce an antibody (immunoglobulin). The antigens have certain areas on the surface, known as **epitopes** which bind to the immunoglobulin. In this process, T-cells participate and present the antigen to the B-cells. The B-cells thus stimulated, proliferate and mature into **plasma cells** that produce specific antibodies. The antibodies bind to the antigens and inactivate them. Some of the activated B-cells become **memory cells** that retain the capacity to rapidly produce further plasma cells on subsequent contact with antigens. Hence, a quicker and greater response to subsequent infection occurs, when compared to primary attack by a microorganism or virus. This is known as **immunological memory**. Plasma cells are short-lived while **memory cells** are long-lived.

The T-cells that help to stimulate and present antigens to B-cells are known as **T-helper cells (TH cells)**. Some TH cells also interact with mononuclear phagocytes and help them to destroy pathogens. Some T-cells are able to destroy bacteria and viruses which have already entered the host cells (intracellular action). They are known as **Tc-cells**. The chemical signals sent by Tc-cells are known as **cytokines**. The recognition of antigens by Tc-cells is because they possess specific antigen receptors. The cell surface has what are known as **MHC proteins**. They are different in unrelated individuals. Hence, T-cells distinguish foreign MHCs as antigens because of the differences in MHC proteins and own MHCs of individuals do not elicit antigen response. T-cells also recognize infected cells because infection alters cell surface substances. Interaction between T-cells and other cells of the immune system are illustrated in Fig. 53.

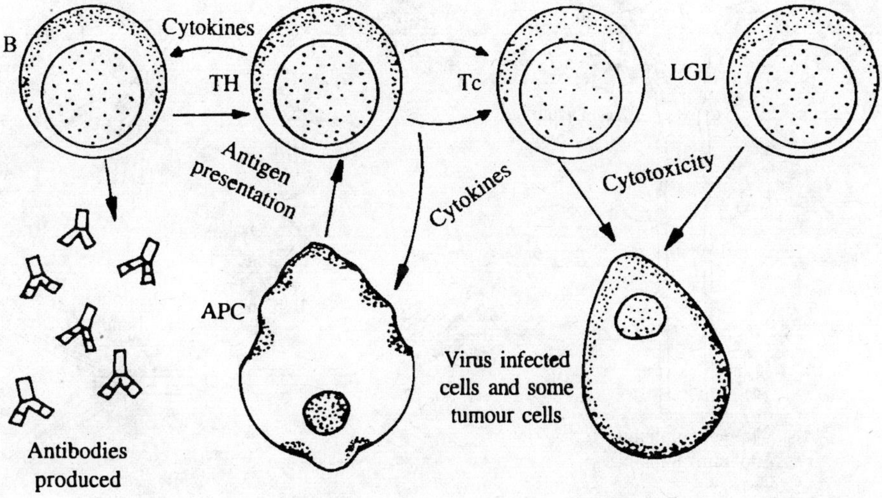

B = B-cell APC = antigen presenting cell
TH and Tc = T-cells LGL = large granular lymphocyte

Fig. 53 Interaction between T cells and other cells of the immune system.

Immunoglobulins

The plasma cells are able to produce specific antibodies because of structural features of immunoglobulins. Five classes of immunoglobulins, namely IgG, IgM, IgA, IgE, and IgD are produced.

Immunoglobulins are soluble molecules with antibody function. The basic structure of an immunoglobulin, known as a **monomer,** is a unit made up of two pairs of polypeptide chains—two identical light chains (L chains) and two identical heavy chains (H chains). Both L and H chains have components of a **variable structure** (V_L and V_H) with NH_2 **terminals,** and a **constant structure** (C_L and C_H) with COOH terminals. The heavy chains have a hinged region linking a first region (CH-1) with a second and a third region, CH-2, and CH-3, respectively.

The L and H regions are linked by disulphide bonds. Within the variable region, there are subregions which exhibit highly variable amino acid sequences. These regions are known as **hypervariable regions** or complementarity determining regions (CDRS). Figure 54 illustrates the basic structure of an immunoglobulin and arrangement of monomers in IgM and IgA.

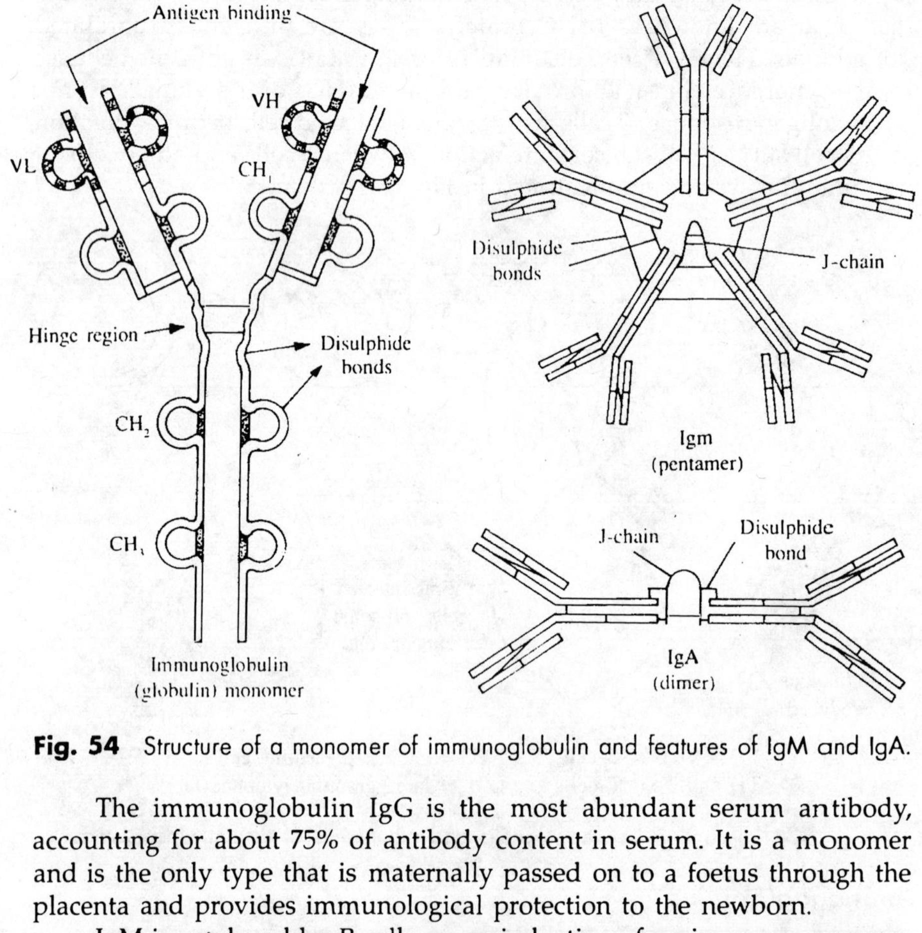

Fig. 54 Structure of a monomer of immunoglobulin and features of IgM and IgA.

The immunoglobulin IgG is the most abundant serum antibody, accounting for about 75% of antibody content in serum. It is a monomer and is the only type that is maternally passed on to a foetus through the placenta and provides immunological protection to the newborn.

IgM is produced by B-cells, upon induction of an immune response. It accounts for 8–10% of the serum antibody content. IgM is a pentamer formed by linking of five monomers by disulphide linkages of C-terminal ends. A polypeptide called **J-chain (joining chain)** also links the five monomers. The IgM antibodies are efficient in binding to bacteria and viruses.

IgA is abundant in certain body secretions such as saliva, tears and secretions of the respiratory and intestinal tracts. It occurs either as a monomer or as a dimer (2 monomers linked together). They provide early defense by binding to pathogens and preventing their adhesion to tissues. They also make the pathogens more susceptible to phagocytic action.

IgE is present in tissues of the spleen, tonsils, adenoids and mucous membranes. Only trace amounts are present in serum. It provides immunity against intestinal parasites. It is involved in allergic responses by releasing histamines from mast cells.

IgD is a monomer, occurring in low concentrations in the serum. Its functions are not clear.

The Igs function as specific antibodies against a wide range of antigens due to the **diversity of variable region** that binds to the antigen. The variable regions of L and H chains are folded in such a way that the hypervariable regions are brought together to create a region that binds the antigen. The regions on the antigen to which antibodies bind are known as **epitopes** and the region on the antibody molecule that binds to the antigen is known as the **fab portion**. The other part interacts with other components of the immune system (neutrophils, macrophages, PMN cells etc.). This part is known as the **FC region**. This binding allows ingestion and digestion of the pathogen by those components. The antibody forms a coating on the pathogen surface (**opsonization**) making possible its recognition by the phagocyte in cellular immunity.

The antibodies produced are very diverse. The number of different forms of the antibody is more than that of all other proteins in the body. It was suggested that constant and variable regions are coded by a single gene, while the HV (hypervariable) region is coded by separate genes. Apart from the presence of several genes, somatic recombination and somatic mutations are responsible for generating antibody diversity. Interaction between light chains and heavy chains produces different specificities. The V regions of antibody are clustered at the end of the Fab arms. Particular sequences at these ends interact specifically with the antigen. The antigen and antibody are bound by formation of multiple non-covalent bonds. Figure 55 illustrates antibody formation by antigen recognition in B-cells, and phagocytosis of a bacterium by a macrophage.

The T-cell antigen receptors also have C and V regions, but they recognize antigens only in the presence of foreign MHC molecules.

Immunity to Microorganisms and Viruses

Bacteria can either invade the host cells and damage them or produce toxins without invading the cells. Some bacteria show both mechanisms of pathogenicity. **Leptomatous leprosy** does not produce toxins while *Clostridium tetani* and *Corynebacterium diphtheriae* produce toxins while several others (*Staphylococcus aureus* etc.) show both mechanisms. The innate immune system of the body recognizes certain bacterial components and removes them from the tissues without specific mechanisms coming into play. **Activation of complement proteins, release of cytokines and T-cell and B-cell mediated antibodies result in the elimination of the pathogens.**

Antibodies play an important role in dealing with bacterial toxins. The diphtheria toxin is neutralized by blocking of its attachment to target cells, or cell mobility is impaired by binding to flagella. Antibodies trigger

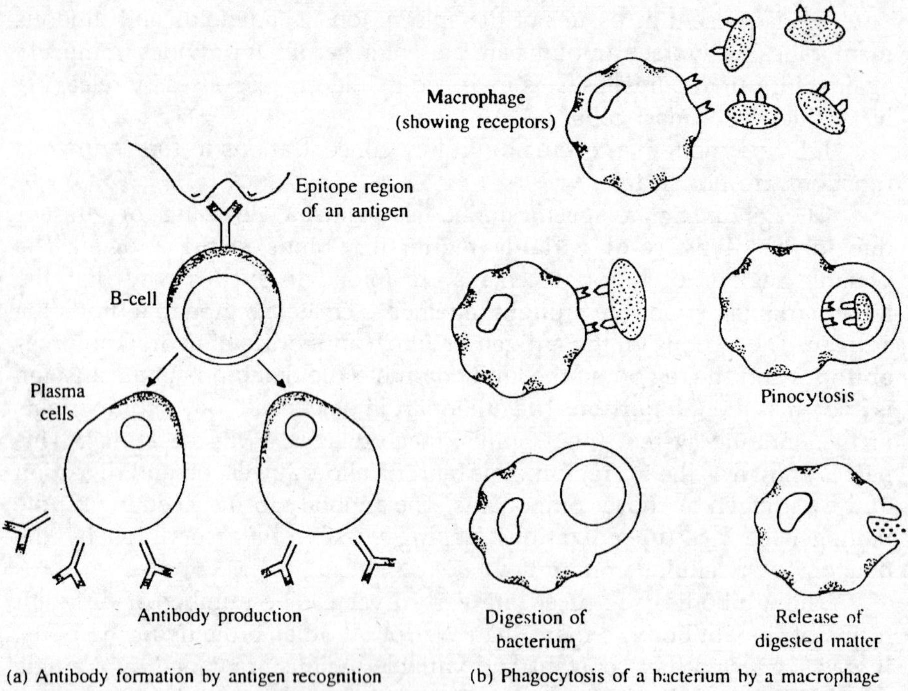

Fig. 55 (a) Production of antibodies by B-cells; (b) Phagocytosis of a bacterium by a macrophage.

complement-mediated damage to the outer membrane of Gram-negative bacteria. Antibodies to M-proteins and capsules bring about opsonization, thus facilitating phagocytosis. Some bacteria like *E. coli*, *Salmonella* and *Pseudomonas* have developed methods of avoiding complement-mediated damage. Table 15 lists immunity mechanism of some important bacterial infections.

Table 15 Immunity Mechanisms in Some Important Bacterial Infections

Organism	*Pathogenic factor*	*Defence mechanisms*
Corynebacterium diphtheriae	Pharyngitis toxin (non-invasive)	Neutralization by antibody
Vibrio cholerae	Enterotoxin (non-invasive)	Neutralization and blocking adhesion
Staphylococcus aureus	Locally invasive and toxic	Opsonized by antibody and complement triggered killing by phagocytes
Mycobacterium tuberculosis	Invasive (evokes immunopathogenesis)	Cytokines from T-cells and macrophage activation

Phagocytes ultimately take part in destroying bacteria after they are bound. The killing mechanism is either **oxgen-dependent** or **oxygen-independent**. In the former type, an enzyme on phagocyte surface reduces the oxygen to superoxide, which is highly toxic. Formation of nitric oxide (NO), which is highly toxic to bacteria and tumour cells also occurs. The oxygen-independent mechanisms include formation of cysteine and arginine-rich cationic peptides which damage lipid bilayers of plasma membrane. Exposure to bacterial products or to cytokines released by lymphocytes activates macrophages to initiate phagocytosis. NK cells and Tc-cells can directly kill Gram-negative bacteria. However, bacteria have developed methods of avoiding phagocytosis. Some release repellents which act against chemotaxis, and others have capsules that inhibit attachment, thus avoiding binding to macrophages.

Immunity mechanisms against fungi are not well understood but they are believed to be similar to those against bacteria. Skin infections are usually self-inhibitory due to the development of resistance to reinfection by cell mediated immunity. It is presumed that TH cells release cytokines which activate macrophages to destroy fungi. Cationic proteins are implicated in immunity to respiratory tract mycoses.

Immunity to viruses. Viruses cause a wide range of diseases. As they lack their own metabolism, they must enter host cells for their replication. They are checked by several immune mechanisms, mostly based on recognition of viral antigens by T-cells. Both external molecules on the virion and its internal components are antigenic. The antibodies formed alter the interaction of virus and host cell which interferes with binding, penetration, uncoating or replication. Tc-cells are involved in immunity to certain viruses.

Interferons are several unrelated proteins released by host cells in response to viral infections and are inhibitory to them. Several mechanisms are involved in their antiviral effects. Direct inhibition of viral replication, activation of NK cells, and increase in MHC proteins which facilitate viral antigen recognition are known.

The HIV (Human immuno deficiency virus) infection results in prolonged latency when it is taken up by T-cells (and also other cells of the immune system). This is due to its integration as a provirus into the host DNA, without any transcription occurring. Certain factors lead to activation of its transcription resulting in increased production of virus-infected cells. Elimination of virus has not been possible due to latency and progressive immunodeficiency caused by T-cell damage etc.

Immunity to Protozoan and Helminth Parasites

Some protozoan parasites like *Plasmodium* (malarial parasite) are small enough to enter host cells while helminths generally occur in gut or lymphatic tissues. Several effector mechanisms have evolved in hosts but the parasites also developed evasive strategies. For instance, *Trypanosoma brucci* (causing sleeping sickness) evokes antibody production and complement stimulation and is inhibited by oxygen-dependent immunity. However, it is able to impair oxygen metabolites and it also evades and resists digestion in macrophages.

ACQUIRED IMMUNITY

Naturally acquired active immunity is mediated by antibody production during pathogenesis. It may be permanent, as for measles or chicken pox, or lasting for only a few years as with tetanus. **Naturally acquired passive immunity** is due to transfer of antibodies from the mother to foetus through the placenta (e.g. IgG).

Artificially acquired active immunity is brought about by immunization with vaccines. When pathogenicity is due to a single toxin, as in diphtheria or tetanus, a toxoid is developed from heat-killed toxin which brings about detoxification while retaining immunogenicity. For pathogens occurring in many serotypes (each serotype being antigenic) it is difficult to develop effective vaccines. Sometimes it is necessary to develop site-specific vaccines. One approach is to develop stable mutants which infect and stimulate immune responses but die, as with *Salmonella* mutants which are used as oral vaccine. Vaccines are designed to elicit immune responses and are applied as a prophylactic measure prior to infection. Attenuated virus particles or bacterial cells, which are rendered nonpathogenic by heat treatment are frequently employed as vaccines. Certain cellular components, especially those associated with cell walls are also used as vaccines. Table 16 lists vaccines for some important bacteria and viruses.

Table 16 Vaccines in Use Against Important Bacteria and Viruses

Disease	*Vaccine*	*Application*
A: Virus diseases		
Influenza	Inactivated virus or components	Yearly
Poliomyelitis	Attenuated (oral vaccine)	Children 2–3 years
Mumps, measles and rubella	MMR combined vaccine (attenuated viruses)	Children 15–19 months
Rabies	Inactivated virus	For individuals in contact with susceptible animals
B: Bacterial diseases		
Cholera	Vibrio cell fraction (immunity lasts 6 months)	During epidemics
Tetanus, Diphtheria and Pertussis	Combined DPT vaccine (toxoid-based) (immunity for 10 years)	Children 2–3 years
Typhoid	Killed cells of *S. typhi* (immunity for 3-4 years	During epidemics
Pneumonia	Polysaccharide from *S. pneumoniae* (capsule)	For adults with chronic infections

HYPERSENSITIVITY AND ALLERGIES

Hypersensitivity or allergy is a body reaction following exposure to certain antigens resulting in exaggerated or inappropriate form of immune response causing tissue damage. Four types of hypersensitive reactions are recognized, types I, II and III being antibody-mediated and causing immediate hypersensitive reactions. IgE when directed against environmental antigens such as pollen, dust, insect parts etc. produces acute reactions such as **asthma, eczema** and **anaphylactic shock**. Delayed hypersensitivity, the fourth type develops 12 hours or more after contact with the antigen. TH 2 type cells and also IgG are involved in this reaction. Eczema, fever, and hardening of the skin are usual symptoms.

When the immune system functions inappropriately, **autoimmune diseases** occur. Several such diseases are known. Pernicious anemia, rheumatoid arthritis, insulin-dependent diabetes etc. are some of the serious autoimmune diseases.

SEROLOGY

The *in vitro* study of antigen-antibody reactions is known as **serology**. Such a study is helpful in determining antigenic relations of isolates of bacteria or of viruses. Antibody reactions are employed in recognizing blood groups, i.e. antigenic groups A, B, AB and O and the presence or absence of Rh factor (Rh positive or negative). This information is vital in blood transfusion.

When an antigenic organism or particle is introduced into the blood of an animal (such as rabbit) antibodies develop in it. The blood is collected and the serum separated. Techniques are available to study its relation to the antigen. Such an antibody-containing serum is referred to as an **antiserum**. When the two are mixed they form aggregates due to attachment (at various epitope regions). The results of such aggregations are visualized and assessed. Traditionally prepared antisera contain more than one type of antibodies because the cell or particle introduced into the blood generally has more than one antigen. Such antibodies are called **polyclonal antibodies**. Their specificity is low. Recently, **hybridoma technique** has been developed to prepare highly specific antibodies known as **monoclonal antibodies**. A specific antigenic substance is introduced into an animal. B-cells are isolated from a suitable tissue of the animal (such as spleen). They are hybridized with tumour cells (myeloma cells) and multiplied in tissue culture. This is because B-cells do not multiply easily whereas myeloma cells multiply continuously. Such hybridoma cells are cultured and monoclonal antibodies are collected in sufficient quantities from the hybrid cells. Such monoclonal antibodies have been very useful in determining taxonomic relations of bacteria and viruses and in diagnosis of diseases.

The following methods are in general use in studies of antigen-antibody reactions in vitro.

Neutralization. When an antitoxin combines with the toxin, the latter is neutralized. Inactivation of certain viruses also occurs when the antibody combines with virus coat protein and prevents its attachment to the host cell.

Precipitation. Antigens and antibodies combine in groups, as they have more than one site of attachment and this results in precipitation of the complex. Several techniques are used to observe the precipitation reaction. For instance, the antigen and antibody are placed in small wells at some distance from each other on an agar plate. Both diffuse in agar and where they meet, an opaque precipitate appears. The technique can be used to compare antigenic affinity of related substances (such as from different strains of a bacterium). If the antibody to one of them is placed in the control well and the antigens are placed around it, formation of precepitate indicates relationship.

Agglutination. For soluble or fluid substances, a carrier support is used for exposure to antiserum. A clumping reaction occurs which can be observed either visually or under a microscope. **Blood grouping** is done by this technique employing antibodies to **A**, **B**, **AB** and **O** groups. Antibodies to **Rh** factor are used to find out if the blood is **Rh positive** or **Rh negative**. The technique of employing blood to observe agglutination reactions is known as **haemagglutination**. Very low concentrations of antibodies can be detected by this technique.

Fluorescent antibodies. In this technique, antisera are first conjugated with a fluorescent dye (e.g. acridine orange). The antisera retain the active sites where attachment to antigen occurs (epitope regions). The cells under study are treated with the conjugated antibodies. The presence of antigen is located by the fluorescence of the dye, which is observed under a fluorescence microscope. This technique has been very useful in locating bacterial cells, virus particles and other antigens in or on host cells.

ELISA (Enzyme linked immunosorbant assay). This is the most frequently employed serological technique now, as it is highly sensitive, rapid and accurate. A transparent plate with a large number of wells is used as a base. It is sensitized by adding antigen in the wells. Excess antigen is washed. Antibody is then added to the wells and excess is again washed. If there is serological relation, the antibody sticks to the antigen. The plate is again washed and an enzyme which gets attached to the antibody is added. A chromogen (which gives colour to the enzyme) is then added and colour is developed. The intensity of colour is measured using a densitometer. The optical density is a measure of the affinity between the antigen and the antibody employed, as well as the antibody content of the antiserum. ELISA is routinely employed to test bacterial and viral antigens, as a large number of samples can be tested simultaneously in the plate. Nowadays, the presence of HIV (which causes AIDS) is tested by this technique.

Immunoelectrophoresis. This is a very sensitive method for detecting individual antigens in a mixture of antigens present in antiserum (polyclonal type). Electrophoresis on a gel is employed to separate individual substances, which get spread on the gel (due to differences in their isoelectric points). The gel is cut into narrow strips across. The strips are placed parallel to each other in a plate. A trough is cut parallel to the strips and the antibody preparation is placed in the trough. Diffusion occurs and precipitate lines form when the antigen-antibody reaction occurs. Figure 56 illustrates some of the serological methods.

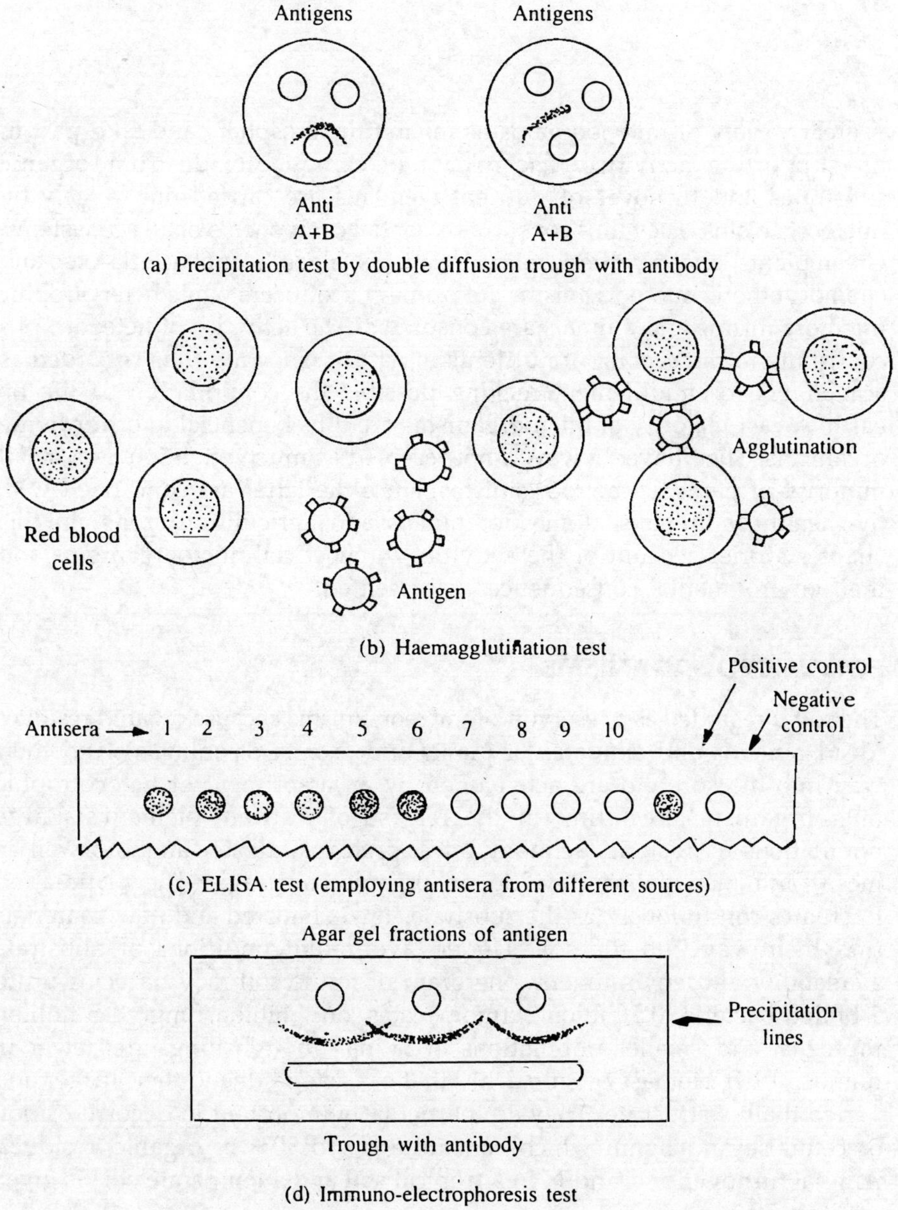

Fig. 56 In vitro methods of studying antigen-antibody reactions.

9

MICROORGANISMS IN THE ENVIRONMENT

A great variety of microoorganisms inhabit the biosphere and take part in life-supporting activities. The release of carbon dioxide from organic substrates and turnover of nutrient elements are carried out largely by microorganisms. They thus constitute essential components of all ecosystems. Green plants, photosynthetic microorganisms (algae, cyanobacteria etc.) and chemosynthetic microorganisms are primary producers while heterotrophic microorganisms and animals are consumers. The activities of heterotrophs contribute to the cycling of nutrients in nature. They are thus regarded as **scavengers** (or middlemen) making possible the continuation of life on earth. A vast majority of microorganisms are thus beneficial to other living organisms. They have, however, not received as much attention as a small minority of parasitic microorganisms, since the latter are concerned with the health of humans, domestic animals and agricultural crops. In this chapter a brief account of the activities of beneficial microorganisms and their environmental consequences is presented.

SOIL MICROORGANISMS

The soil is regarded as a reservoir of microorganisms because organic residues (dead remains and exudates of plants and excreta of animals) find their way into the soil and are acted upon by an assortment of heterotrophic microorganisms, resulting in the release of nutrient elements. Large populations of bacteria, actinomycetes, protozoa, algae, fungi, as well as meso- and macrofauna develop on organic substrates. The population fluctuates continuously, as the substrates are exhausted and new materials find their way into the soil. Under favourable conditions of substrate availability and environment, one gram of fertile soil may harbour up to 2 billion bacteria, 0.5 billion actinomycetes, one million fungi, 0.5 million protozoa, and smaller populations of fauna. Though fungi are fewer in number, their biomass often reaches that of bacteria due to their larger and varied thallus structure. They are particularly important in decomposition of cellulose and lignin which constitute up to 90% of organic residues. Annual turnover of C and N in a tropical soil and a temperate soil is given in Table 17.

Table 17 Annual Turnover of Carbon and Nitrogen in a Temperate and a Tropical Soil

Determinant	*Temperate soil (U.K.)*	*Tropical soil (Brazil)*
Soil weight (tonnes/hectare)	2200.00	2400.00
Organic carbon (tonnes/hectare)	26.00	26.00
Carbon inputs (tonnes/hectare/year)	1.20	13.00
Carbon turnover (years)	22.00	2.00
Microbial carbon (tonnes/hectare)	0.57	0.46
Microbial nitrogen (tonnes/hectare)	0.10	0.08
Microbial turnover (years)	2.50	0.24
Nitrogen flux (kg/hectare/year)	34.00	350.00

Biogeochemical Cycles

The activities of microorganisms and their interaction with the physical environment result in cycling of nutrient elements in nature. Heterotrophs contribute to these changes in large measure, but chemoautotrophs also bring about certain transformations that contribute to the processes. Their activities contribute to maintaining fertility of the soil. Humus is a highly resistant organic constituent of soil. It imparts desirable properties to soil texture and moisture retention and also acts as a long-term reservoir of nutrients for heterotrophs. Microbial activities also contribute to humus formation. Figure 57 illustrates the release of nutrient elements and formation of humus by soil microbial activities.

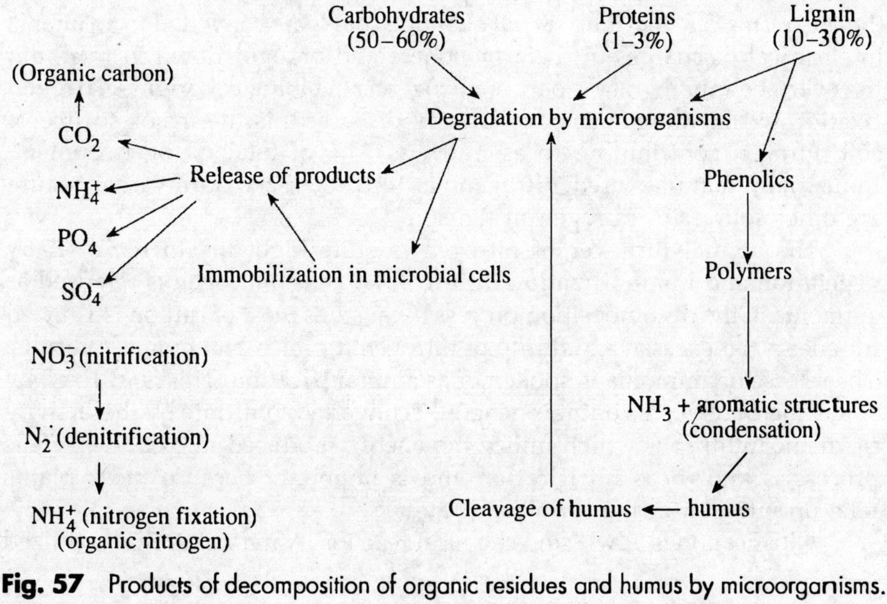

Fig. 57 Products of decomposition of organic residues and humus by microorganisms.

Carbon cycle. The most important contribution of microorganisms is the release of **carbon dioxide** into the atmosphere during decomposition of organic matter. Under aerobic conditions, 60–90% of carbon in organic residues is utilized as source of energy, releasing CO_2 in respiration. Under anaerobic conditions, as much as 95–98% is released. A highly reduced form of carbon, **methane** (CH_4) is formed under anaerobic conditions with rich organic residues. It is a highly energy yielding substance and is utilized by some groups of microorganisms as explained earlier. Thus, microbial activities are vital in **global cycling of carbon**, as illustrated in Fig. 58. Besides microbial activities, burning of fossil fuels (coal and petroleum) and respiration in plants and animals are minor sources of CO_2 in the atmosphere. Carbonates present in rocks and sediments are also contributory.

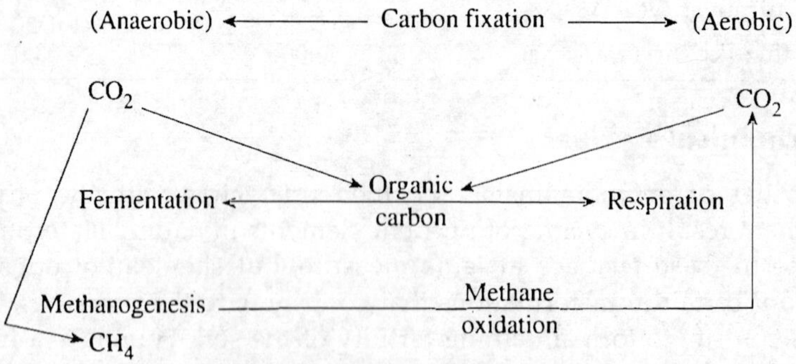

Fig. 58 Carbon cycle in nature.

Nitrogen cycle. Various transformations of nitrogenous substances in soil are largely the function of microorganisms. Nitrogen gas, though most abundant in the atmosphere, cannot be directly used and has to be converted to a reduced form, involving the input of large quantities of energy before the organisms can take it up. Small quantities of N_2 are converted to ammonia in electrical discharges in the atmosphere and brought down in rain, but the overwhelmingly major part of nitrogen is of biological origin. **Nitrogen fixation** by microorganisms, as already discussed, is the major source of soil nitrogen contributing to as much as 75% of total nitrogen content. Industrially manufactured **nitrogenous fertilizers** and **farmyard manure** are other sources of nitrogen in the soil.

The annual turnover of nitrogen is quite high due to **removal by vegetation and immobilization in microbial cells.** Nitrogen is released as ammonia in the decomposition process because, as most of carbon is utilized for energy, an excessive build-up of nitrogen for microbial biomass results. The release of ammonia is spoken of as **ammonification**. This is utilized by some microorganisms but it is generally converted to **nitrate** by the activity of chemoautotrophs which utilize the energy produced in oxidation. This process is known as **nitrification** and is important because most plants take up nitrogen in the form of nitrate.

Nitrification is a two-step chemical reaction. Ammonia is first oxidized to nitrite by *Nitrosomonas* and *Nitrosococcus*. This step requires molecular oxygen as electron acceptor. Nitrite is a toxic substance but it is quickly oxidized to nitrate by another set of organisms, of the genus *Nitrobacter*. The nitrifiers require molecular oxygen and are inhibited under anaerobic conditions or when heterotrophs are very active, reducing oxygen availability. Hence, soil aeration is an important factor in nitrification. The biochemical steps in nitrification are illustrated below:

Ammonia $\longrightarrow$ hydroxylamine $\longrightarrow$ nitrite

$NH_3 \longrightarrow (NH_2OH) \longrightarrow NO_2^-$ (*Nitrosomonas* etc.)

Nitrite $\longrightarrow$ nitrate

$NO_2^- \longrightarrow NO_3^-$ (*Nitrobacter*)

Under anaerobic conditions, nitrate is converted to molecular nitrogen and is lost into the atmosphere. When organic residues are abundant, microbial activity is high and oxygen gets depleted. The microbes then utilize nitrate as **terminal electron acceptor**. This is in effect **nitrogen respiration**, and is known as **denitrification**. The steps in the conversion of nitrate to molecular nitrogen are as follows:

$$NO_3^- \longrightarrow NO \longrightarrow N_2O \longrightarrow N_2$$

Nitrate	Nitric oxide	Nitrous oxide	Nitrogen

Besides depletion of nitrogen by plant removal and immobilization, nitrogen is lost due to denitrification. Small quantities of nitrates are also lost by **leaching** to lower layers of soil. These losses are made up by nitrogen fixation. Figure 59 illustrates the nitrogen cycle.

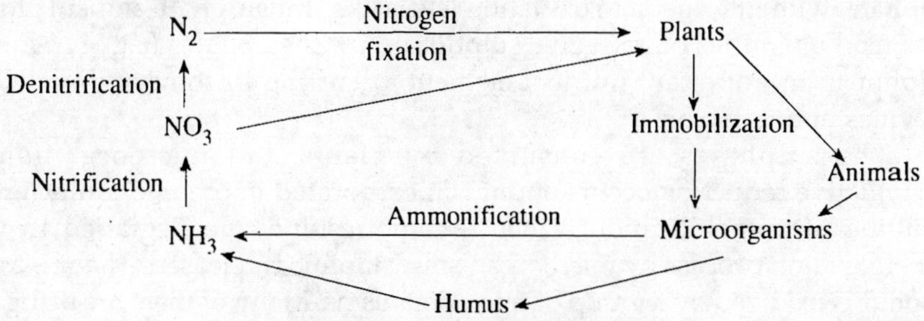

Fig. 59 The nitrogen cycle.

Phosphorus cycle. Phosphorus is the third most important element needed by organisms. Soil organisms are involved in **solubilization** of phosphorus compounds, which are otherwise not available to plants. They are said to be in the fixed form of phosphorus. They occur mainly as apatite (rock phosphates). Phosphorus fertilizers contain phosphoric acid, which is a soluble form of phosphorus. Chemical reactions in soil keep an equilibrium between soluble and insoluble forms of phosphorus, thus making it necessary for phosphatic fertilizers to be regularly added for adequate supply to crop plants.

Biologically, phosphorus occurs in a bound form, mainly as **phytin** (inositol hexaphosphate), which is synthesized both by plants and microorganisms. **Phospholipids,** and **nucleic acids** are the other sources of biological origin. Phytin is decomposed very slowly while the other forms are degraded rapidly by microorganisms.

Many microorganisms solubilize phosphorus compounds. Fungi such as *Aspergillus* species produce organic acids which cause solubilization. Some common bacteria such as *Pseudomonas* and *Bacillus* show similar capacity. *B. megaterium* was found to be efficient and is employed as a **biofertilizer,** under the commercial name of *phosphobacterin*. **Vesicular-arbuscular mycorrhizal fungi,** which are endosymbionts in plant roots, extend into soil from roots and have been shown to be important in supplying phosphorus to host plants. Figure 60 illustrates the biological transformations of phosphorus in soil.

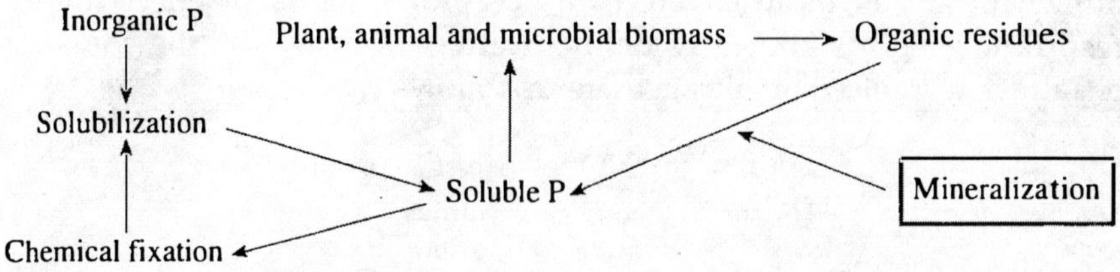

Fig. 60 The phosphorus cycle in soil.

Sulphur cycle. Sulphur is another macronutrient whose transformations in soil are largely biological in nature. In soil it occurs mainly as an organic fraction. Its oxido-reduction states range from H_2S, **elemental sulphur** to **sulphate,** with intermediate oxidation levels like **thiosulphate** and **sulphite,** etc. Small quantities of inorganic sulphur occur as sulphates (e.g. gypsum). Sulphur is an important nutrient element, occurring in some amino acids, enzymes and cofactors.

The sulphates are consumed by plants and microorganisms. Assimilative reduction occurs and it is incorporated in organic compounds resulting in immobilization. When organic residues are degraded in the mineralization process by microorganisms, sulphur is released. The released sulphur is oxidized by a group of microorganisms. Many of them are **obligate lithotrophs (chemoautotrophs),** utilizing the oxidation energy for their life processes. *Thiobacillus thiooxidans* and other species oxidize elemental sulphur or reduced compounds forming sulphates, as outlined below:

$$SH^- \longrightarrow S \longrightarrow S_2O_3 \longrightarrow S_4O_6 \longrightarrow SO_4^{2-}$$

Sulphide ⟶ Elemental ⟶ Thiosulphate ⟶ Tetrathionate ⟶ Sulphate

Elemental sulphur, or gypsum (calcium sulphate) is applied to correct the pH of alkaline soils as the sulphuric acid that is formed lowers it.

$$\underset{\text{Sulphur}}{2S} + 3O_2 + 2H_2O \longrightarrow \underset{\text{Sulphuric acid}}{2H_2SO_4}$$

Thiobacillus thioparus and T. ferrooxidans are also **obligate lithotrophs.** The latter oxidizes iron or sulphur compounds and is implicated in corrosion of iron pipes in soil. *T. novellus* is a **facultative autotroph** growing either on organic or inorganic sulphur compounds. *T. denitrificans* grows as an autotroph under aerobic conditions and as a denitrifying organism under anaerobic conditions, oxidizing sulphur while reducing nitrate to molecular nitrogen.

$$5S + 6KNO_3 + 2H_2O \longrightarrow 3K_2SO_4 + 2KHSO_4 + 3N_2$$

Colourless filamentous bacteria (*Beggiatoa, Thiothrix*) and photosynthetic sulphur bacteria also oxidize sulphur compounds.

Sulphates are reduced under anaerobic conditions rich in organic compounds, such as lake sediments. Two obligate anaerobes, *Desulfovibrio* and *Desulfomaculatum* carry out sulphur reduction using SO_4 as electron acceptor. The sulphur cycle is illustrated in Fig. 61.

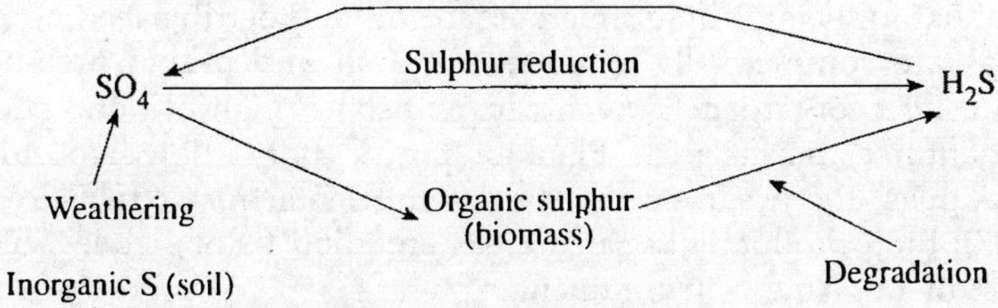

Fig. 61 The sulphur cycle in soil.

Other transformations. Iron is another element that is subjected to microbial activities. Micronutrients like zinc, copper, molybdenum, manganese and boron are also involved in microbial transformations.

Biomining

Bacteria are now being used for **leaching** out minerals from mines. They are in regular use for leaching **copper** and **uranium** from ores. Zinc, nickel cobalt, lead and gold were also shown to be leached by bacteria. The most important organisms are *Thiobacillus ferrooxidans, T. thiooxidans, Leptospirillum ferrooxidans* and species of **sulpholobus**. All of them have autotrophic metabolism, oxidizing inorganic substrates. The following reactions bring out their activities.

$$\underset{\text{(Iron pyrite)}}{4FeS_2} + 15O_2 + 2H_2O \longrightarrow \underset{\text{(Ferric sulphate)}}{2Fe_2(SO_4)_3} + 2H_2SO_4$$

$$CuS + Fe_2(SO_4)_3 \longrightarrow CuSO_4 + 2FeSO_4 + S$$

(Covellite) + Ferric sulphate ⟶ Copper sulphate + Ferrous sulphate + S

$$UO_2 + Fe_2(SO_4)_3 \longrightarrow UO_2SO_4 + 2FeSO_4$$

(Uranium oxide) ⟶ (Uranium sulphate)

MICROORGANISMS IN AQUATIC HABITATS

The diverse aquatic habitats like oceans, estuaries, rivers, lakes and ponds show considerable variation in physical and chemical properties, and so harbour diverse population. The depth of a body of water exhibits changes in light, temperature and aeration. The surface layer is **hypolymnion,** below which **thermocline** occurs. There is a gradual reduction in light, temperature and oxygen tension in the thermocline. The lower region presents more stable conditons and is known as the **benthic zone**. The organisms floating on the surface or suspended in upper layers are mostly **algae** and **cyanobacteria**, constituting **phytoplankton** which are photosynthetic, accounting for primary production of organic matter. Zooplankton and other fauna develop resulting in complex communities.

Areas near the shore and estuaries are rich in communities as the nutrients are brought in from rivers and streams. Variations in salinity result in a diversity of organisms. Deposition of organic residues provides a good habitat for heterotrophic microorganisms and detritus feeding fauna. Such areas are commercially exploited for **fish and prawn production**. **Halophilic microorganisms** which can withstand high osmotic pressure are abundant in estuarine areas. Plankton productivity is low in oceans but in certain areas, the ocean current brings nutrients from coastal regions, resulting in high productivity. Such areas are good fishing zones, like the west coast of the American continent.

The lower, or benthic zone of oceans and lakes also harbours microbial and animal communities, which are based on organic sediments. The productivity is generally very low. As light does not penetrate those areas and temperature is low, only specialized microorganisms develop. Low temperature tolerant **psychrophiles** and those withstanding high barometric pressure occur in those areas of estuarine environment.

Hot sulphur springs which occur in certain areas are habitated by **thermophilic microorganisms**. In recent times an unusual phenomenon was discovered. There are **hydrothermal vents** in the ocean floor where springs of hot water emerge. Such areas were found to be rich in animal communities. Sulphur oxidizing bacteria are found and it is believed that organic matter accumulated by these chemolithotrophs is the source of food for the animal communities.

MICROORGANISMS AND POLLUTION

In intensive farming practices, the use of **xenobiotic chemicals** such as pesticides and herbicides is widespread and has become a matter of environmental concern. Pesticide residues find their way into soil and water and gradually get concentrated in the food chain, causing damage and death to consuming organisms. Some of the extensively applied pesticides of the past are now banned because of long periods of persistence in soil and water and their resistance to attack by microorganisms. These chemicals, known as recalcitrant chemicals such as DDT, last for several years as do organochlorine pesticides like Aldrin, Heptachlor etc. Therefore, their use is being discontinued and xenobiotics which are **biodegradable** and disappear from the environment are now favoured. The half-life (period of exposure required to reduce the original chemical concentration to 50%) of commonly employed organophosphatic pesticides like monocrotophos is only one to two weeks. Herbicides like 2, 4-D are also biodegraded in a short time, while the related 2, 4, 5-T persists up to one year in soil. Several heterotrophic bacteria have been isolated from soil which can metabolize some pesticides. The common soil bacteria which degrade pesticides are species of *Bacillus, Pseudomonas, Flavobacterium, Mycobacterium* and *Nocardia*.

Worldwide use of petrochemicals has greatly increased pollution of soil and water by hydrocarbons. A few microorganisms, notably *Pseudomonas, Micrococcus* and *Corynebacterium* were found to metabolize those compounds. Some yeasts (e.g. *Torulopsis*) can grow on hydrocarbons. Attempts are being made to utilize them for production of single cell proteins. Detergents and effluents from the plastics industry are pollutants of water bodies. Polychlorinated biphenyls in effluents are highly resistant and much attention is being paid to isolate microorganisms that can metabolize them. Detergents which are biodegradable are now favoured. *Pseudomonas fluorescens* is known to metabolize such detergents. Several methanogenic bacteria such as *Methanobacterium, Methanococcus, Methanosarcina, Methanomicrobium, Methanospirillum* are active.

MICROORGANISMS IN SEWAGE

The vast amounts of domestic sewage and industrial effluents are treated for purifying and recycling of water for use. The sewage is very rich in nutrients and a variety of microorganisms develop. One consequence is rapid depletion of oxygen leading to anaerobic conditions. The need for oxygen supply under these conditions is referred to as **biological oxygen demand** (BOD). It is measured in closed bottles of sewage samples, incubated at 20°C for a fixed period. Initial and final readings of dissolved oxygen are the measure of BOD.

The aim of sewage treatment is to bring down the BOD to acceptable levels. Several processes are employed to increase oxygen supply so that microbial activity continues and organic matter is decomposed as completely as possible. The traditional engineering methods employed in sewage treatment plants in urban centres are the **activated sludge process** with surface aeration by rotating the sewage in large and shallow tanks, or by forcing air through the sewage. Another process is the use of **trickling filters** where the liquid is made to trickle through solid particles, so that liquid and air are in continuous contact. The efficiency of sewage treatment is increased by selecting efficient microorganisms and adding them to treatment systems, as **starter cultures.** *Pseudomonas putida* is one such organism patented for this purpose. Another process that is becoming popular is **anaerobic digestion** of sewage or industrial effluents (such as distillary effluents) which are rich in nutrients. Methane, an important energy source is recoverd in the anaerobic process.

METHANE PRODUCTION

Methane is a clean-burning fuel and one of the rich energy sources. To tap this source, anaerobic digesters are designed for utilizing agricultural and industrial wastes. Organic substrates are digested to initially form CO_2 and H_2. These are reduced under anaerobic conditions to form methane (CH_4). The entry of air to the digester is restricted so that oxidation is prevented. The methane is channelled to burn as fuel. **Gobar gas** produced from dung is one of the extensively used fuels produced in anaerobic digesters. Sugar factories use molasses to produce methane as the energy source. Several methanogenic bacteria such as *Methanobacterium, Methanococcus, Methanosarcina, Methanomicrobium, Methanospirillum* are active.

10

MICROORGANISMS IN INDUSTRY AND FOOD

The varied fermentative metabolism of microorganisms has been an important factor in food processing and pharmaceutical industries. The traditional processes of production of milk products, alcoholic beverages, bread, and vinegar are due to microbial activity, which has been employed by mankind for centuries, although the role of microorganisms was not understood in early times. Estimates of major fermentative products in 1981 show alcoholic beverages taking the first place, accounting for 23 billion, followed by cheese at 14 billion, antibiotics, 4.5 billion and high fructose syrups at about one billion pounds sterling. Several industrial chemicals such as citric acid, amylases, lactic acid, dried yeast, and monosodium glutamate (a flavouring salt) are major industrial products besides industrial alcohol and acetic acid that are produced by microbial fermentation.

In the early period, most of the organic chemicals were produced by fermentation but their share in production declined after the discovery of petroleum-based chemical methods of production. Ethylene was employed as a starting point for synthesizing a range of organic chemicals. However, with increases in petroleum costs, some of the fermentation products became economical and are now produced by this route. Moreover, food additives such as flavouring agents, colours, emulsifiers, thickening agents, etc. are not allowed to be produced from petrochemicals and so fermentation products are used. Agricultural products rich in sugars and starch are common feed stocks for fermentation. Although cellulose is the most abundant raw material, methods for its use have not been economical due to its high resistance to breakdown.

Corn, an extensively produced cereal grain rich in starch, has been used to produce a range of fermentation products. Every component of grain is used in industry. Wet milling of corn produces powdered starch, corn syrup, corn oil, gluten and corn steep liquor. The latter product has been an important raw material for the production by large scale fermentation of antibiotics and other pharmaceutical chemicals.

Starch is first subjected to **enzymatic hydrolysis** with alpha amylases and glucoamylases. About 98% of starch is converted to glucose. It is then fermented with yeast (*Saccharomyces cerevisiae*) for ethanol production. Carbon dioxide is produced as a by-product and is bottled and used in carbonated beverages which improves their taste (aerated water, soft drinks etc.). Figure 62 gives the flowchart of fermentation products starting from glucose.

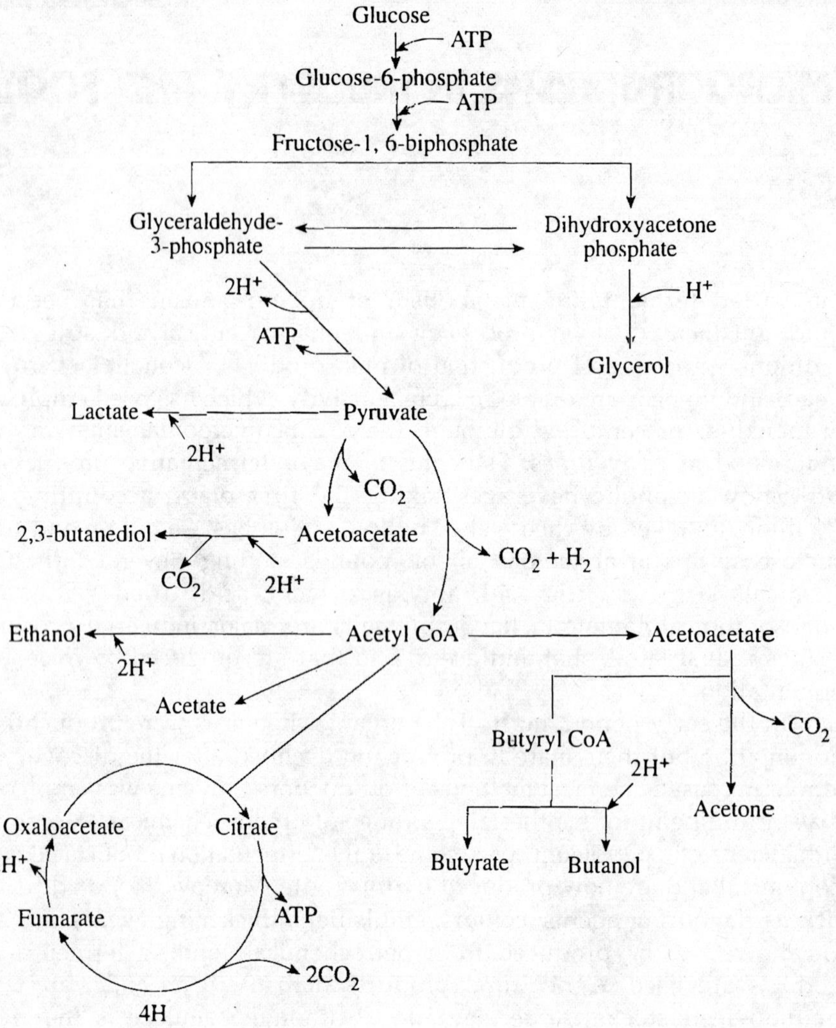

Fig. 62 Production of industrial chemicals through fermentation, starting from glucose (glycolysis and citric acid cycle are shown partially).

Fermentation with other microorganisms produces a number of industrially important chemicals. Table 18 lists important fermentation products, organisms employed, and principal uses.

RAW MATERIALS FOR FERMENTATION

Raw material costs, pretreatment of raw materials, additives such as enzymes, cost of maintaining inoculum, cost of materials and processes for extraction and purification of the product, have all to be considered.

Table 18 Important Substances Produced through Microbial Fermentations

Product group	*Product*	*Producing microorganisms*	*Principal uses*
End products of energy metabolism	Ethanol (food)	*Saccharomyces cerevisiae*	Beer, wine, CO_2 and yeast biomass
	Ethanol (industrial)	*S. cerevisiae and Kluyveromyces fragilis*	Base for chemicals, automobile fuel
	Acetic acid (vinegar)	*Acetobacter spp.*	Food acidifier
	- do -	- do -	Base for chemicals
	Lactic acid	*Lactobacillus delbruckii, L. bulgasicus etc.*	Textile and leather industries
	n-Butanol	*Clostridium acetobutylicum*	Solvent
	2, 3-Butandiol	*Aerobacter aerogenes*	Base for synthetic rubber
Enzymes and proteins	α-amylase	*Bacillus licheniformis, B. subtilis*	Starch hydrolysis
	β-amylase	*B. polymyxa, Rhizopus niveus*	- do -
	Taka diastase	*Aspergillus niger*	Digestive enzyme
	Insulin	*Genetically engineered E. coli*	Hormone for diabetes treatment
	Interferon	- do -	Antiviral and anticancer protein
	Lysine	*Corynebacterium glutamicum*	Food supplement
	Single cell protein	*Candida utilis*, Methylotrophs	Food and feed suppliment
Metabolic intermediates	Citric acid	*Aspergillus niger, A. wentii*	Food and industry
	Fumaric acid	*Rhizopus nigricans*	Resins, wetting agents
	Glutamic acid	*Brevibacterium flavum*	Flavour enhancer in foods
	Vitamin B12	*Propionibacterium shermanii*	Correction of anemia
Secondary metabolites	Gibberellins	*Fusarium moniliforme*	Plant growth hormone
	Penicillin	*Penicillium chrysogenum*	Antibiotic
	Streptomycin	*Streptomyces griseus*	- do -
	Tetracyclines	*S. aureofaciens, S. rimosus*	- do -
	Erythromycin	*S. erythreus*	- do -
	Cephalosporin	*Cephalosporium acremonium*	- do -
	Rifamycin	*Nocardia mediterranea*	- do -
	Gentamicin	*Micromonospora purpurea*	- do -
	Polymyxin	*Bacillus polymyxa*	- do -
	Griseofulvin	*Penicillium griseofulvum*	- do -
	Hydroxyprogesterone	*Rhizopus arrhizus*	Hormone

Abundantly available agricultural products such as corn grain, soybean powder and malted barley or industrial wastes such as molasses from the sugar industry, sulphite wastes from the paper industry and whey (spent liquid in cheese-making) are commonly employed. Corn steep liquor, a product of corn wet milling industry is a frequently used source for nitrogen and vitamins. For fermentations which have to be more exactly controlled, defined substances such as glucose, sucrose, and other sugars, or starch are employed. Starch from manihot tubers is a common substrate. Starch is first hydrolyzed by amylases to release glucose which is then fermented. Beer production is based on maltose, which is formed by malting of barley. Cellulose is the most abundant plant material available but because of its high resistance to breakdown, efforts to utilize it as a substrate in commercial fermentations have not been successful although it is being done experimentally. Milk is the raw material for cheese and other fermented dairy products. Fruit juices are the starting point for wines and spirits and for food grade vinegar.

FERMENTATION PROCESSES

In the early days, fermentation technology was simple. Anaerobic fermentation was carried out in large vats or open wooden vessels to produce alcoholic beverages or solvents. With the expansion of food processing and pharmaceutical industries based on microorganisms, elaborate systems have been evolved. Fermentation vessels were designed to maintain conditions of sterility, aeration, pH, viscosity and temperature at optimum levels required for each system.

The cost of a fermented product depends on several factors such as cost of raw materials, cost of additives like enzymes and vitamins, cost of running the fermentation system, maintaining the inoculum, maintaining sterility of the medium, control of temperature, aeration and viscosity of the medium and time taken for the fermentation cycle. Constant check must be maintained on the quality of the inoculum because contamination by unwanted microorganisms, or alteration by mutations, reduces yields and complicates recovery of the products. To achieve economies, a 3-4 day cycle is employed and vessel size is determined for optimum performance. Where costly enzymes are used in the process, immobilization of enzymes in suitable carriers is employed so as to recycle them for repeated use.

Three types of processes are employed in large-scale fermentations.

Batch fermentation. This is a closed system in which all the nutrients are initially added to the vessel and inoculated. Subsequent treatments include maintaining adequate aeration by providing stirrers and pH control by the addition of an acid or alkali. In aerated systems, antifoam agents such as palm oil or soybean oil are added. Large-scale growth of microorganisms tends to generate heat in the system. Temperature control is maintained by providing water circulation jackets around the vessel for heat exchange.

In large fermentations, the **fed-batch method** is employed to control the concentration of critical nutrients. For instance, penicillin production is controlled by maintaining low glucose level in the medium as its increase causes catabolic repression of secondary metabolites (most antibiotics are secondary metabolites).

Continuous fermentation. This is an open system in which the nutrient solution is kept in a reservoir and continuously fed into the vessel. An equal amount of fermented nutrient medium with the microorganism is withdrawn continuously. This maintains the nutrient composition constant and allows continuous growth of the organisms. Such a fermentor is called a **chemostat**. Arrangements are made to maintain sterility of the medium and also sterility of air where aeration is provided. Viscosity of the medium is kept low because dissolved oxygen content reduces as viscosity increases. Microbial biomass and nutrient level remain more or less constant in a chemostat. In a variation of the system, microbial growth is assessed on turbidity of the medium. Such a system is referred to as **turbidostat.**

Plug-flow method. This is also a continuous fermentation system but here a tall cylinder is employed as the fermentor, instead of a large vessel. A mixture of nutrient medium and inoculum are made to trickle through the cylinder and the exuding fermented medium and inoculum are collected at the outflow.

Figure 63 illustrates the three types of fermentation systems.

Specific growth rates of different organisms depend on a number of factors such as limitation of substrate concentration, cell density and yield constants. Conditions for maintaining maximum specific growth are selected and a suitable system is employed. For instance, in case of fungi, since they are aerobic an adequate supply of oxygen should be ensured whereas strict anaerobic conditions should be maintained for anaerobic organisms like *Clostridium.* Continuous systems are employed in fermentation of antibiotics, organic solvents and single cell protein (SCP).

The stages of cell growth and product formation may be simultaneous or separated in time. In one type of fermentation, the product is directly derived from primary metabolism for energy yield. Hence, cell growth and product formation occur simultaneously, as in ethanol or SCP production. In a second type, cell growth occurs initially but product accumulation occurs after growth slows down. Citric acid and some amino acids accumulate this way. In a third category, the products are secondary metabolites and accumulate at a later stage of growth but cell growth continues during that phase also. Many antibiotics are formed this way. Selection of fermentation process and the length of the fermentation cycle take these aspects into consideration.

Product Recovery

Purification and recovery of the desired product is a critical aspect of fermentation. Several steps are usually involved in separation of the product from substances in the fermentation broth, and from unwanted metabolites formed during fermentation. The recovery process should maintain the stability of the product. The first step in recovery is separation of solids and liquids by sedimentation or filtration. For large batches, filter aids such as **diatomaceous earth** are used. Large-scale centrifuges are also used in some systems. Very often the desired product is in low concentration in

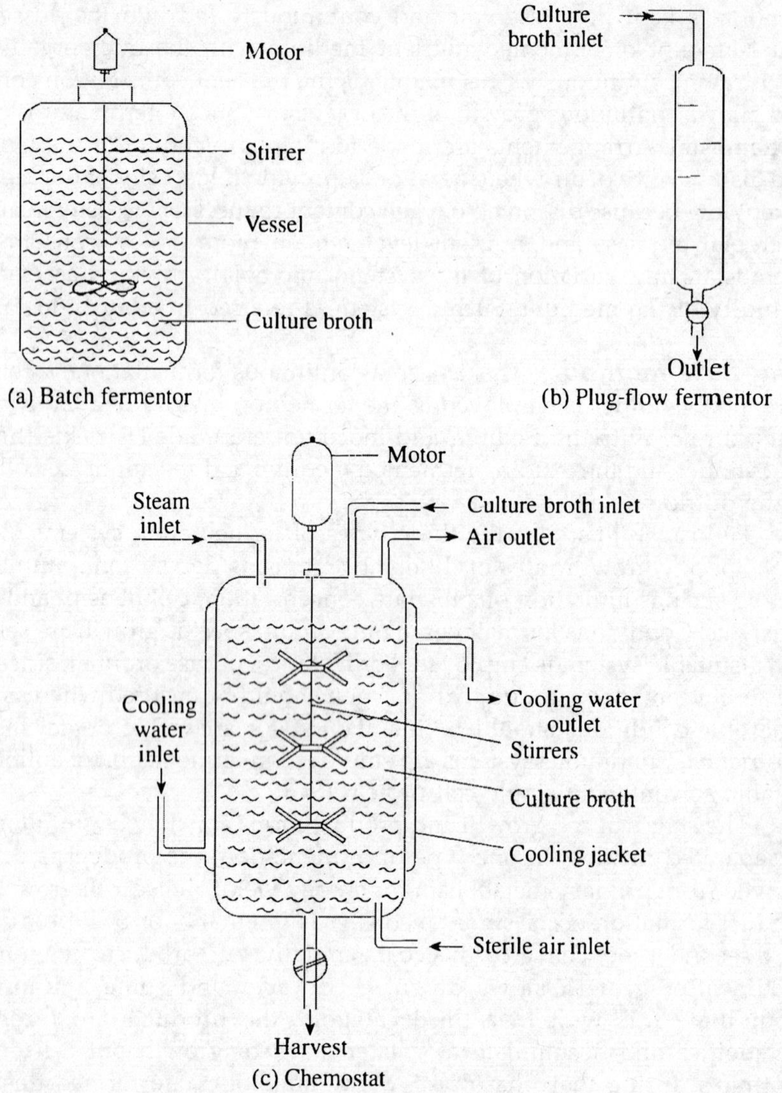

Fig. 63 Three main types of fermentation vessels.

the filtrate. The filtrate (or culture broth) is concentrated prior to extraction and purification so as to reduce the quantity of solvents used. Rotary film evaporators, membrane filters, ion-exchange resins and counter current distribution, etc. are employed for concentration, extraction and purification of the desired product. Fat-soluble substances are extracted in solvents like **n-butanol** or **ethyl acetate**. Simple distillation is sufficient for ethanol and other volatile substances.

The products are released into the culture broth in some fermentations, retained in the cells (intracellular) or found in both states in others. Disruption of cells by physical or chemical methods like ultrasound, ball mills, osmotic solutions etc. is carried out prior to extraction in suitable solvents. The process of purification depends on the degree of purity required for the end use. For instance, products for medical use have to be highly pure and may involve elaborate purification procedures. Maximum recovery and stability of the products also have to be ensured. Microbial whole cells are the desired products as in the case of bakers' yeast, single cell protein (SCP) etc. They are separated from other ingredients, dried and stored.

Strain Development

The success of commercial fermentations of industrial chemicals depends on several factors that influence economies. Except in the food industry (alcoholic beverages, cheese and other milk products etc.) where naturally-occurring strains of microorganisms are employed, few of the organisms originally isolated from nature are in use. Production of antibiotics, enzymes, amino acids and other products from improved strains is achieved through improvement programmes such as induction of mutants, genetic recombinations and more recently through recombinant DNA techniques. Mutants are selected with the aim of increasing the quantity of primary product, decreasing the fermentation time, reducing the formation of undesirable products (which reduce the quantity or interfere with recovery and purification of primary product), tolerance increasing to low oxygen tension and growth on inexpensive substrates. Table 19 lists the increases achieved by strain improvement.

Table 19 Increase in Antibiotic Production Achieved through Strain Improvement

Product	*Productivity (units/ml)*	
	At the time of discovery	*After strain improvement*
Penicillin	20	8000
Streptomycin	50	5000
Erythromycin	100	2000
Oxytetracycline	400	6000

The mutagenic agents generally employed are physical agents such as short wavelength UV radiation (200–300 nm, with optimum at 254 nm, which is the maximum DNA absorption wavelength), ionizing radiations including *X*-rays, β-rays, γ-rays and neutron or proton bombardments, and chemical mutagens such as DNA base analogs (5-bromouracil, 2-aminopurine etc.), alkylating agents (ethyl methane sulphonate, nitrosoguanidine etc.) or those causing direct damage to DNA (hydroxylamine). The parasexual phenomenon was employed for achieving genetic recombination in fungi (*Penicillium, Cephalosporium, Aspergillus, Claviceps*). Conjugation has been employed in *Escherichia coli* and *Pseudomonas spp.* and transformation in *Bacillus spp*. Conjugation and protoplast fusion were achieved in actinomycetes. Strain improvement by biochemical approaches has also been made. Excess formation of enzymes, vitamins and amino acids was achieved by eliminating feedback mechanisms, end product regulation and catabolic regulation.

Genetic engineering methods have been successfully developed for production of a few substances of importance in medicine. Transfer of human genes into strains of *Escherichia coli* by recombinant DNA technology and recovery of gene product from culture is now in use for the production of insulin (for diabetes treatment), interferon (antitumour and antiviral agent) and somatostatin (human growth hormone to reverse dwarfism).

PRODUCTS OF INDUSTRIAL FERMENTATIONS

The following categories which are produced through microbial activity are discussed citing examples of important products in each category. Many antibiotics are formed this way. Selection of the process and the d uration of the fermentation cycle are made taking these aspects into consideration.

Ethanol. Ethanol is produced from raw materials such as **molasses**, a by-product of the sugar industry, **starch** from corn or *Mannihot esculenta* tubers and, in the case of alcoholic beverages, from **malted barley** (for beer), **grapes** (for wine) or other sugary fruits. Bakers or brewers yeast, *Saccharomyces cerevisiae* is the organism usually employed but for production of ethanol for industrial uses, another yeast, *Kluyveromyces fragilis* is sometimes used. Ethanol production is done by the batch process.

Yeasts are aerobic organisms and the process is initially maintained under aerobic conditions to form adequate biomass (yeast cells). Anaerobic conditions are then created resulting in stoppage of growth, accumulation of ethanol, and CO_2 evolution. Yeasts tolerate ethanol concentration of 12–18% in the medium after which production stops completely. The CO_2 is collected and bottled and used as aerating gas. When starch is used as a substrate, it is first converted to glucose by **pretreatment with amylases**. Ethanol is separated by a simple process of distillation. Distillation results in collection of 95% ethanol (rectified spirit) which is used as a starting material for synthesis of certain chemicals in industry. The yeast biomass is separated and used as animal feed. Strains of yeast which produce less ethanol but more CO_2 are used for bread making etc. (Bakers yeast). These days, ethanol is used as an alternate fuel for automobiles. Rectified spirit is further distilled over benzene to form 100% ethanol. It is used either directly as fuel or in mixture with petrol. Since enthanol is produced from renewable plant sources and is ecofriendly (clean burning), its use as fuel is encouraged. Brazil, which has a large sugar industry is trying out ethanol as alternate fuel.

Acetone-butanol fermentation. This is an anaerobic process employing the obligate anaerobe *Clostridium*. A mixture of acetone, butanol and ethanol, in a ratio of 30 : 60 : 10 is produced when C. *acetobutylicum* is employed and only butanol is formed when C. *butylicum* is employed. Acetone and butanol are industrial solvents.

Molasses and **corn steep liquor** media are employed, as they are rich in sugars, proteins, and growth factors required for these anaerobic organisms. Glycolysis followed by incomplete oxidation of acetyl CoA results in the accumulation of these solvents. Carbon dioxide and hydrogen gases and ethanol are recovered as by-products.

Vinegar and acetic acid. Acetic acid is one of the organic acids traditionally produced by fermentation. Vinegar, which is an acidulant employed in food preparations, is produced from fermented fruit juices. Acetic acid is also produced for use as a chemical intermediate in industry. It is produced in two stages. **Fruit juices** are first fermented with yeast to form **ethanol** (an anaerobic process as described earlier). It is then trickled through a specially designed vessel so that adequate aeration is provided for oxidation of ethanol to **acetic acid** by the action of *Acetobacter spp.*, which are aerobic. Acetic acid for industrial uses is prepared, starting from ethanol. The chemical equation is as follows:

$$2CH_3CH_2OH + 2O_2 \longrightarrow 2CH_3COOH + 2H_2O$$

The fermented product (or ethanol) is recycled several times through the system until a desired concentration of acetic acid is obtained. The system has to be maintained at 15–34°C and aeration has to be adequate.

Citric acid. Citric acid is a major fermentation product and has several uses as a flavouring agent in foods as an antioxidant and preservative, and as a component of detergents. Annual world production of citric acid is about 250000 tonnes. It is produced in aerobic fermentation systems employing fungi such as *Aspergillus, Penicillium, Mucor, Paecilomyces* and *Candida*. Molasses is the usual raw material. *Aspergillus niger* is commercially employed on a medium maintained at pH 5.0 to 6.0. Iron content in the medium must be kept at a low level. The medium is clarified and citric acid is precipitated as a calcium salt by adding lime to the clarified medium.

Lactic acid. Lactic acid is produced either by fermentation or by chemical means, but the former route is employed when it is used in the food industry. The raw materials commonly employed are **whey** and **sulphite liquors**. *Lactobacillus delbruckii, L. bulgaricus* and *L. pentosus* are used and the anaerobic process is employed because *lactobacilli* are facultative anaerobes and anaerobic conditions prevent complete oxidation of glucose to CO_2. The microorganisms employed show **homofermentative pathway** of lactic acid production, which is desirable.

Gluconic acid. Gluconic acid is used in metal and leather industries, sodium gluconate in detergents and calcium gluconate in medicine. It is formed from glucose through the action of glucose oxidase. In commercial production, *Acetobacter suboxidans* and *Aspergillus niger* are employed. Immobilized enzyme systems have been adopted for its production.

Amino acids. Several amino acids are used in the food industry and manufacture of medicines to correct nutrition deficiencies. **Glutamic acid** is the base for the production of sodium glutamate, an extensively used flavour enhancer in foods. Glutamic acid production is estimated at 3,00,000 tonnes annually. L-glutamic acid is produced by fermentation with *Corynebacterium glutamicum.* Glucose, sucrose or molasses are the commonly employed substrates, and **biotin** is added to the medium. Best yields on molasses are upto 100 g/litre. Other amino acids are also produced by fermentation but commercially extracted from protein hydrolysates (as in the case of lysine, methionine etc.). **Lysine** is also produced by fermentation, by a two-step process employing *Escherichia coli* on a molasses medium, which forms **diaminopimelic acid (DAP)**. The fermentation broth is then incubated with *Aerobactor aerogenes* for 20 hours at 35°C. The DAP is converted to L-lysine.

Enzymes. A number of enzymes used in industry or medicine are produced by microorganisms. Starch is a commonly used substrate. It is a branched chain polymer of glucose and must be hydrolysed prior to its use as a substrate (**saccharification**). α-amylases, β-amylases and isoamylases are needed for complete saccharification. α-amylases are produced by several species of *Bacillus. Bacillus licheniformis, B. amyloliquifaciens* and *Aspergillus oryzae* are used for fermentation employing the batch method. β-amylases are usually derived from plant sources though microorganisms produce them. **Isoamylases** which split the α-1, 6-bonds of starch (at branch points in the chain) are produced by *Pseudomonas amylodesmosa*. Starch is made into a paste and subjected to enzyme action at 55°C. The resulting starch solution is used as substrate.

Glucose isomerase, which converts glucose to fructose is employed for the production of a syrup. It is a mixture of glucose and fructose and is used in fruit beverages and confectionary (it is sweeter than sucrose or glucose). *Bacillus coagulans* and *Streptomyces albus* are employed for glucose isomerase production.

Proteases are extensively used in detergents and in cheese making. Several species of *Bacillus* and *Streptomyces* produce alkaline proteases for use in detergents. *Bacillus cereus* is employed for fermentation in an alkali medium (pH8-9). **Rennin**, a milk coagulating enzyme is used in cheesemaking. It is derived either from intestines of calves or produced by microbial fermentation. *Mucor pusillus* and *Endothia parasitica* are commercially employed for the purpose.

Pectinases are mainly used to clarify and to reduce viscosity of fruit juices. *Aspergillus niger* or *A. wentii* are employed in fermentation.

Penicillin acylase is employed for the breakdown of penicillin G to form **6-amino penicillanic acid,** from which many semisynthetic penicillins are made. Penicillin acylase is produced by *Escherichia coli* in a nutrient medium consisting of 2% corn steep liquor. Genetically engineered strains have been developed for increasing yields.

Immobilization of enzymes. Purified enzymes lose their effectiveness during storage. Hence, they have to be **stabilized**. Moreover, they are expensive and ways must be devised to use them repeatedly to achieve economic viability. Different methods of stabilization for various enzymes have been developed. Enzyme molecules may be cross-linked to each other in such a way that their activity is not affected. Enzymes may be stabilized with carrier substances or encapsulated in semipermeable membranes, which permit their activity to continue. These methods are referred to as **immobilization** of enzymes.

Several enzymes including α-amylases, alcohol dehydrogenases and proteinases are cross-linked with a polymerizing agent. **Glutaraldehyde** is commonly employed. Binding the enzyme to a carrier is the commercially employed method. **Bentonite**, activated carbon, starch, tannin etc. are employed as carriers. Ion exchange resins are also employed. Covalent bonding is achieved with substances such as **CMC (carboxy methyl cellulose)** which is also commonly employed.

Immobilization is also achieved by enclosing the enzymes in water-insoluble gels. **Polyacrylamide gel** is frequently employed. Penicillin acylase, glucose isomerase, lactase etc. are commercially employed as immobilized systems.

Vitamins. Vitamin B12 (5-deoxyadenosyl-cobalamino phosphate) is the only vitamin commercially produced through fermentation, though fermentation methods were previously employed for riboflavin etc. Glucose is the substrate and *Pseudomonas denitrificans* the most efficient producer of vitamin B12. Present yields are 60 mg/litre on a molasses medium supplemented with yeast extract and minerals. *Propionibacterium* species also forms vitamin B12 in fermentation.

Antibiotics. Apart from that of ethanol and alcoholic beverages, production of antibiotics is the most important aspect of industrial microbiology. Antibiotics are **secondary metabolites** formed by a large number of organisms, especially soil microorganisms. They are inhibitory to other microorganisms. They are characterized by selective action. Hence, some of them have proved highly useful in controlling infectious diseases. Their discovery and development have truly revolutionized treatment of very severe diseases.

Chemically, antibiotics represent different groups and microorganisms producing them include fungi, actinomycetes and bacteria. Table 17 lists important antibiotics in clinical use and the organisms from which they are isolated. The basic chemical structures of different chemical groups are illustrated in Fig. 64. The following account discusses fermentation aspects of those antibiotics.

β-Lactum antibiotics. **Penicillin** and **cephalosporin** belong to this group. 6-aminopenicillanic acid possesses similar molecular structure, with

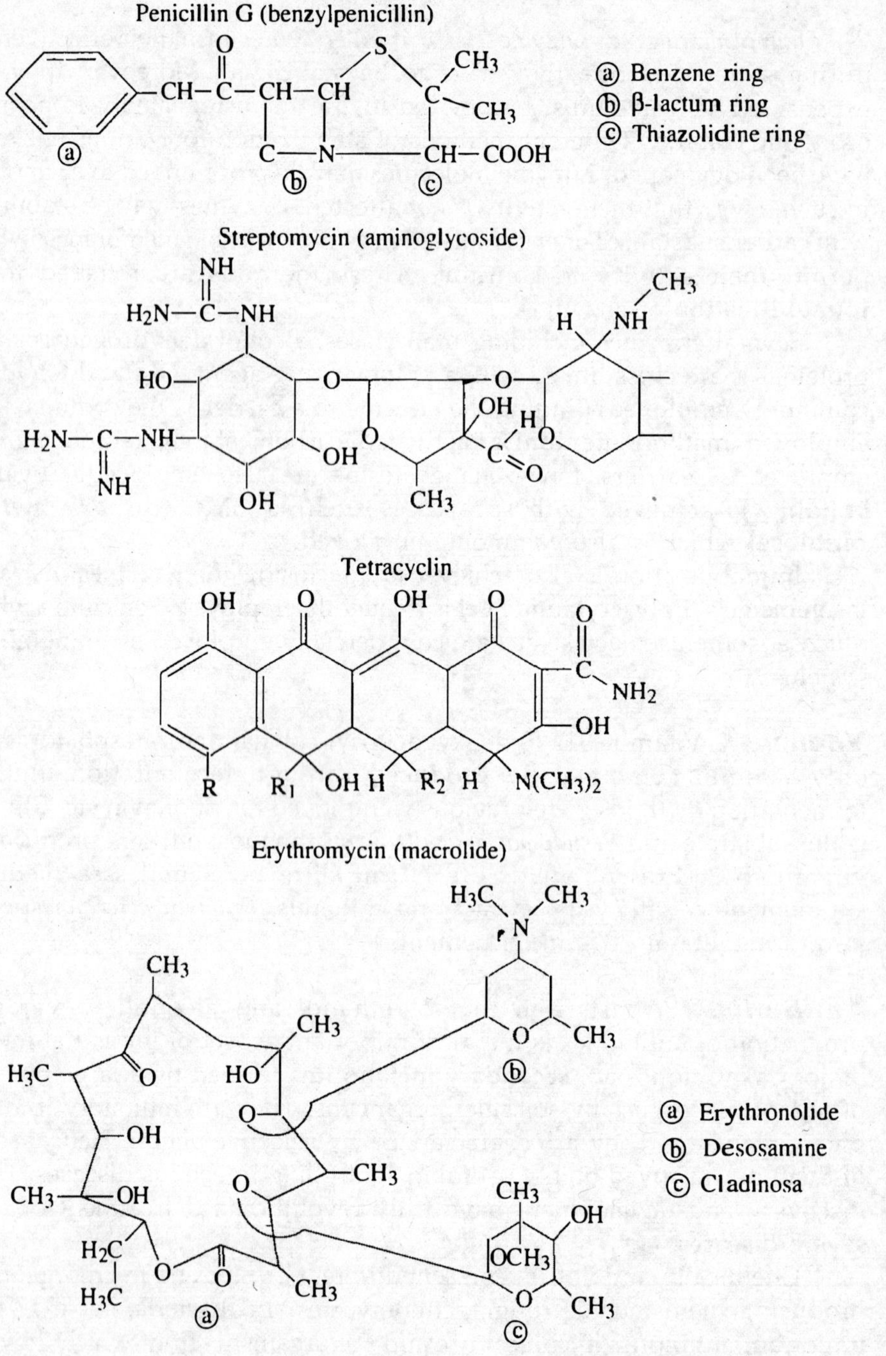

Fig. 64 Basic chemical structures of main groups of antibiotics.

a β-lactum and thiazolidin ring structure. Penicillin G is produced in large-scale submerged fermentations, employing improved strains of *Penicillium chrysogenum* in the batch process. From this 6-aminopenicillanic acid is extracted with penicillin acylase. Semisynthetic penicillins, with better clinical properties are prepared from the basic molecule thus prepared. The fermentation medium is made up of corn steep liquor, soymeal, yeast extract and whey, pH 6.5. Penicillin is extracted from fermented broth and is purified by the **counter-current method**.

Cephalosporin was originally isolated from culture filtrates of *Cephalosporium acremonium*. An improved strain is now used in fermentation, by the same method as for penicillin. Many semisynthetic cephalosporins are in clinical use. Cephalosporin has a broader spectrum of activity (larger groups of microorganisms are inhibited) and a very low toxicity. Hence, it is one of the most frequently employed drugs. The medium is made up of corn steep liquor, glucose and ammonium acetate, at pH 7.0. The growth phase lasts for 90 hours followed by a production phase stretching for upto 160 hours.

Peptide antibiotics. *Bacitracin* produced by *Bacillus licheniformis* and *polymyxin* produced by *B. polymyxa* are in topical use for external application (powders and ointments) because of their toxicity. Fermentation occurs in aerated submerged vessels and solvent extraction with *n*-butanol is used for separation.

Aminoglycosides. **Streptomycin, Neomycin, Gentamicin** and **Kanamycin** belong to this group. Their basic structures are made up of oligosaccharides with aminocyclianol moiety attached to them. Neomycin is produced by *Streptomyces fradiae,* kanamycin by *S. kanamyceticus* and gentamicin by *Micromonospora purpurea*. They are in therapeutic use, especially against Gram-negative pathogenic bacteria.

Streptomycin, which is the first antibiotic shown to be active against tuberculosis, is produced by *S. griseus*. It is fermented on a medium of soymeal and glucose in aerated, submerged cultures and extracted from the medium by the ion exchange methods.

Macrolides. **Erythromycin** and **rifamycin** are important antibiotics in this group. They consist of **12–17 lactone rings** connected through deoxyhexose sugars. Erythromycin has 14 lactone rings. It is effective against Gram-positive bacteria and is used as an alternative to penicillin. It is produced from fermentation by *Streptomyces erythraeus* in an aerated submerged system on a complex culture medium containing soymeal and glucose. Rifamycin is effective against Gram-positive bacteria and more importantly *Mycobacterium tuberculosis*. It is a **macrolide** with an aromatic component. It is produced by *Nocardia mediterranea*.

Tetracyclines. There are several tetracyclines produced by species of *Streptomyces*. The first tetracycline isolated was chlorotetracycline from *S. aureofaciens*. (It was isolated in Lederle Laboratories of which Dr. Yellapragada Subba Rao was the director.) Oxytetracycline is isolated from *S. rimosus*. The basic molecule of tetracyclines is a **naphthacene ring structure**. Tetracyclines are bacteriostatic (suppressing bacterial multiplication) and have a very wide spectrum of activity, inhibiting Gram-positive and Gram-negative bacteria, rickettsiae and mycoplasmas. About six semisynthetic tetracyclines are in clinical use. Tetracyclines are produced in large, stirred fermentors on a glucose medium with a limitation of phosphate supply.

Single cell protein. The term **single cell protein (SCP)** refers to **microbial biomass used as food** or feed supplement. In recent years, the cyanobacterium, *Spirulina* has become prominent as a source of SCP. The yeasts, *Candida tropicalis* and *Saccharomycopsis lipolytica* are currently employed for mass production of cells. They have the capacity to utilize long-chain hydrocarbons (petroleum wastes are utilized). However, with increases in petroleum product costs, SCP production is now more expensive. *Candida tropicalis* is produced on a fuel oil in a continuous process with forced aeration. **Methanol** is also employed as substrate for SCP production for growing *Methylomonas methalonica* and *Pseudomonas methylotrophus*.

Gibberellins. These are plant hormones influencing plant growth and are extensively employed in fruit and vegetable production. They are produced by the fungus *Fusarium moniliforme* (conidial stage of *Gibberella fujikuroi*). A low-nitrogen medium with a mixture of sugars is the fermentation medium in a batch culture with aeration.

Steroids. Steroids are natural hormones such as *progesterone* and *estrogens*. Substances derived from steroids, especially **cortisone** and **dehydrocortisone** have found therapeutic use. Their production and modification to form the desired types of corticosteroids have been difficult and expensive. Microorganisms were found to be capable of producing the modifications. A two-step process involving microorganisms had been evolved to produce them. Cortisone is produced by chemical synthesis and is then subjected to fermentation by specific microorganisms which carry out precise modifications. The modified substance is extracted and purified. A number of steroids are produced this way. *Hydrocortisone*, for instance, is formed by *Curvularia lunata* and *predinosolone* by *Arthrobacter simplex*, on media containing steroids.

MILK PRODUCTS

Milk is an excellent medium for many microorganisms. Even as milking is carried out, microorganisms in the udder get into the milk. Estimates of fresh milk indicate population of a few hundred to several thousand organisms per millilitre. On storage, a sour smell develops due to increase in lactic acid bacteria, especially *Streptococcus lactis* and *S. cremoris*, which utilize lactose in milk. Other microorganisms like *Escherichia coli* and *Clostridium* develop causing spoilage of milk, due to degradation of proteins and lipids which are abundant in milk.

In order to preserve milk, **pasteurization** is carried out at around 63°C for 10 minutes so as to eliminate microorganisms while preserving its nutrient qualities. However, some heat-resistant organisms like *Streptococcus thermophilus* and lactobacilli like *Lactobacillus* and *Leuconostoc* are not killed. Over a period of time, milk products which can be stored longer have been developed. The two important products are **cheese** and **yogurt** (*dahi*). Cheese-making is mostly carried out on a large industrial scale while yogurt is made in individual households, though some industrial scale production is also being done. Flavours and textures of cheese or yogurt are based on the use of strains of lactobacilli which were evolved over a period of time, and these strains are employed so as to maintain the quality.

An average sample of milk has the following composition: water—87.3%, lactose (sugar) 5.0%, butter fat—3.8%, casein (milk protein) 2.5% and minerals and other substances 0.4%; pH 6.7–6.9, and specific gravity 1.035.

Yogurt is prepared by adding a small quantity of starter culture, which is a previously fermented sample, so that the same microorganisms develop and desired flavour is maintained. Sometimes cream is separated from milk and fermented. The product is **sour cream**. The lactobacilli which are generally found in mother cultures are *Lactobacillus acidophilus, L. bulgaricus, Leuconostoc citrovorum, L. mesinteroides, L. dextronicum, Streptococcus lactis* and *S. cremoris*. They are facultative anaerobes and require complex nutrient media and milk is an excellent medium for them. They are acid-tolerant and the lactic acid formed by them in milk inhibits other bacteria by lowering the pH. Hence, fermented milk products do not spoil for considerable periods.

Cheese is an important milk product and is highly nutritious. Milk is coagulated to separate the solids and liquids. **Rennet,** an enzyme extracted from the intestines of calves or produced by fermentation of *Mucor pusillus* or *Endothia parasitica* is employed for separating the solids. The liquid portion, called whey, is a by-product and is often used as a substrate in other fermentations.

The separated solid mass is **ripened** by addition of cultures of lactobacilli. Different types of cheese have their own distinctive flavours, on account of the microbial strains employed. Species of *Streptococcus* and *Lactobacillus* are employed for making cheddar cheese, the most common type. Swiss cheese is made with *Pediococcus* and *Leuconostoc*. Strong flavoured cheeses are made by further ripening with fungi as with **Roquefort** (*Penicillium roqueforti*) and **Camembert** (*P. camembertii*). Fully ripened cheese is pasteurized before packing.

MICROBIOLOGY OF PERISHABLE FOODS

Fruits, vegetables, meat, fish and poultry are described as perishable foods. They have high nutritional value and high moisture content rendering them excellent substrates for a wide range of microorganisms. Handling of these products during harvest, as also packing and transport to consuming centres provides conditions suitable for the growth of microorganisms. Vegetables and fruits may be harbouring surface microorganisms and pathogens. Damage caused by insects, and mechnical damage due to friction increases the susceptibility to attack by the exposure of live tissues. Precautions must, therefore, be taken to minimize microbial growth on these foods. In the case of foods produced seasonally, methods of long-term preservation have been developed so as to simplify handling and transport and to ensure availability throughout the year.

Vegetables, having high pH and **redox potential** are good substrates for bacterial attack while fruits which are acidic are commonly attacked by fungi because of their acid tolerance. However, microorganisms occurring on the surface as well as those causing rotting of the tissues include moulds, yeasts and bacteria. The frequently occurring bacteria on vegetables are *Pseudomonas, Alcaligenes, Flavobacterium, Lactobacillus, Streptococcus, Serratia* and *Erwinia*. Several fungi such as species of *Aspergillus, Penicillium, Rhizopus, Mucor, Cladosporium, Fusarium*, and *Stemphylium* occur on vegetables and fruits. Several yeasts such as *Candida, Torula, Torulopsis, Rhodotorula* and *Hansenula* also occur. The enzymes produced by these organisms cause rotting of tissues and release of nutrients which are utilized by the microorganisms. Undesirable substances such as slime and odours and discolourations are produced. Visible mould growth can be seen on rotting fruits and vegetables. Toxic metabolites may also be released by bacteria and fungi. Disposal of the spoiled fruits and vegetables also presents problems.

Meat and meat products serve as excellent sources for bacteria on account of their very high nutrient value with high protein content. Meat is also subject to **autolysis** by the **proteolytic enzymes** present in it and the products of autolysis are easily decomposable substrates for microorganisms. The bacteria colonizing meat include *Pseudomonas, Alcaligenes, Leuconostoc, Bacillus* and *Micrococcus*. Pathogenic microorganisms such as *Salmonella, Streptococcus, Staphylococcus* and *Clostridium* also occur.

Spoilage of meat under aerobic conditions (by *Pseudomonas, Alcaligenes, Leuconostoc, Streptococcus* etc.) results in the formation of surface slime. Colour changes are brought about by yeasts, *Flavobacterium* etc., resulting in red or yellow spots. Many **foul smelling** compounds such as **mercaptans** are formed during proteolysis, especially under anaerobic conditions. This is described as **putrefaction**. Oxidation of fatty substances in meat also produces undesirable colours. The more serious problem is the formation of toxins by microorganisms, especially by *Salmonella* and *Clostridium*, which cause food poisoning in improperly cooked meat.

Fish harbour bacteria on their surfaces. *Pseudomonas, Flavobacterium, Corynebacterium, Serratia* and *Bacillus* are common and they are implicated in the formation of surface slime. Besides these organisms, *Vibrio, Clostridium* and *Escherichia* are also found in their intestines. Fish products deteriorate rapidly as they contain abundant enzymes that cause autolysis. That is why they are frozen quickly after they are caught. Spoilage is indicated by fading of natural bright colours, slime formation, pinkish or yellow colouration of gills and softening of tissues. Damage is caused by psychrophilic bacteria like *Pseudomonas, Achromobacter* and *Flavobacterium*. Salted fish are spoiled by osmophilic forms like *Alcaligenes, Micrococcus* and *Serratia*.

Food Preservation

Several methods of food preservation have been evolved. They are based on different principles such as drying, heating, lowering of pH, increasing sugar or salt content and radiation treatment. Most bacteria are killed when they are exposed to temperatures of 55 to 73°C for about 10 minutes. However, some spore-forming bacteria are highly resistant to heat and require temperatures of 120 to 130°C for a few minutes, as with the endospores of *Bacillus* and *Clostridium*. Lowering of pH of food material such as fruit juices reduces the time required for killing the bacteria, thus preserving its nutritional quality.

Pasteurization followed by refrigeration or sweetening of milk is commonly practised. Sterilization of meat and other foods is carried out at temperatures above 100°C, in pressure vessels (autoclaves) where steam at high pressure increases temperature.

Canning is a method of preservation of foods in closed and sealed containers. It is generally preceded by heating and eliminating air at the time of sealing. Properly canned foods can be preserved for years as they exclude contamination by microorganisms. However, if the heating process is incomplete and highly resistant organisms such as *Clostridium* remain, spoilage of canned food occurs. Canned foods may be divided into acidic foods such as vegetables and fruits (tomatoes, citrus fruits etc.) and low acid foods such as meat, fish, poultry, as well as vegetables like peas, beans and corn. The temperature and time of treatment depend on the level of acidity because the lower the pH, the quicker the destruction of microorganisms.

Preservation of foods by increasing osmotic value through sugar syrups, salt, citric acid or vinegar has been an age-old practice. Microbial growth is retarded under these conditions if initially osmophilic microorganisms are not present and the treated materials are properly closed to prevent contamination. A wide range of vegetables and fish are pickled by this method. Many food materials are processed prior to pickling.

Low temperature preservation is employed for some foods. Milk may be chilled and stored for a few days. Many vegetables, fruit juices and fish are frozen and marketed in a frozen state. **Quick freezing devices** have been evolved to prevent microbial development. Temperature is usually lower than – 25°C. Liquid nitrogen is generally employed as the refrigerant. It is claimed that quick freezing preserves the texture and nutritional quality, better than slow freezing, by rapidly slowing down enzymatic and chemical changes in foods.

Drying of foods has been in practice for centuries. Reduction in moisture content retards microbial growth. Moisture content of vegetables and fruits ranges from 75 to 90% while that of grains is 6 to 10%. **Sun drying** can be successfully employed in hot climates with a large number of sunny days. However, during rains and when temperatures are low, evaporation rates would be slow, allowing the development of microorganisms during the drying period, and under those conditions, mechanical driers are employed. A number of devices with forced air circulation such as tunnel trays, rotary drums, and film driers are employed. Milk is generally **spray dried**. Dehydration of foods by **freeze drying** is in operation for high-value foods. **Vacuum drying** is also in use. Table 20 lists drying methods employed for common foods.

Table 20 Types of Drying Employed for Food Materials

Product category	*Type of dryer*
Vegetables, fruits, pectin	Tunnel tray
Grain, apple pulp, distillery wastes	Rotary dryer
Coffee, tea, milk, fruit juices	Spray dryer
Milk, soups, distillery by-products	Rotary film evaporators
Malt extract, coffee essence, meat extracts	Freeze driers or vacuum driers

Food materials are sometimes preserved by using chemical preservatives. Chemicals are also used in conjunction with other methods. Sodium or calcium propionate, sodium benzoate, sodium or potassium sorbate are frequently added to processed foods such as fruit juices, salads, pickles, cheese, carbonated beverages etc. Nitrites and nitrates are employed in meat preservation. Sulphites and sulphur dioxide are used in wine industry, especially for sterilizing equipment. The maximum concentration of chemical preservatives is around 0.1% in food preparations.

Spoilage of Foods

Grain, syrups, pickled foods, fruits etc. are spoiled by **osmophilic organisms** such as *Bacillus* and *Leuconostoc* and certain yeasts (*Saccharomyces, Candida, Rhodotorula*). Fungi such as *Aspergillus, Penicillium, Stemphylium, Cladosporium* are acid-tolerant and also osmophilic to some extent. They cause spoilage of fruit syrups and pickled foods.

Diseased individuals among freshly harvested fruits and vegetables carry surface microorganisms and pathogens. They deteriorate rapidly both from the action of their own autolytic enzymes and of microorganisms. Bacterial soft rots caused by *Erwinia* in cruciferae (cabbage, cauliflower), onions and beet are common. Dry rots are caused by *Alternaria* and *Fusarium* in tomatoes, lemons, potatoes etc. Soft rots in many fruits and vegetables are caused by *Rhizopus*.

Lactobacilli cause spoilage of fruit juices. In concentrated fruit juices, osmophilic yeasts cause spoilage because of their tolerance to high acid and salt content.

In meat, the occurrence of *Salmonella, Clostridium, Streptococci* and *Staphylococci* is common. They form toxins. *Salmonellas* are intestinal pathogens. If meat is not properly preserved, a wide range of bacteria, yeasts and moulds grow on it. Canning is a common method of preservation. Meats are cured for preservation by smoking or treating with brine. Under aerobic conditions, spoilage is by *Pseudomonas, Alcaligenes, Micrococcus* etc. which form surface slime, stickiness and ropiness. Under anaerobic conditions, souring and putrefaction occur, brought on mainly by *Clostridium, Pseudomonas* and *Alcaligenes.*

Spoilage of milk results in gas formation by *Bacillus* and *Clostridium,* **proteolysis** by several bacteria, sour flavour by *Streptococcus* and *Alcaligenes* and ropiness by *Alcaligenes* and *Micrococcus.* Chilled milk is spoiled by psychrophilic strains of *Pseudomonas.*

Canned foods may be spoiled either due to survival of microorganisms during heat treatment or due to leakage in the container permitting the entrance of organisms during storage. In the first situation it is due to the presence of heat-resistant spores of bacteria. When the containers leak, the organisms which find their entry may or may not be heat-resistant.

Formation of hydrogen gas is one of the results of canned food spoilage. Normally the closed ends of cans are flat. When gas is formed, swelling and distortion can be visually observed. Sometimes swelling of the cans is not seen but souring of contents occurs due to lactic acid formation. It is common in low-acid foods like peas and corn. *Bacillus coagulans* which produces acid without gas is the common contaminant. Under anaerobic conditions, *Clostridium thermosaccharolyticum* causes splitting of sugars resulting in CO_2, hydrogen gas formation and souring. *Clostridium nigrificans* forms hydrogen sulphide. It is a thermophile and occurs only if cooling is slow. Many species of *Clostridium* and *Bacillus* which are spore formers occur in lightly heated foods. *C. butylicum* and *C. pasteurianum* produce butyric acid. *C. sporogenes, C. putrifaciens* and *C. butylicum* are proteolytic and cause putrefaction, forming **H_2S, ammonia, indole, skatole** and **mercaptans** all of which are foul smelling. Carbon dioxide and hydrogen may also be produced, causing swelling of cans. *C. botulinum* forms **botulitoxin,** the most potent toxin known and frequent cause of food poisoning deaths. This is known as **botulism**. A few milligrams of the toxin can kill a person by paralysing the muscles. The toxin is, however, thermolabile and is destroyed by heating to 90°C for about 10–15 minutes.

The other forms of food poisoning usually result in gastrointestinal disorders. Contamination with intestinal parasites like *Vibrio, Salmonella* and *Escherichia coli* are responsible. A situation known as **food intoxication** results from contamination with *Bacillus cereus* and *Staphylococcus aureus.*

Mycotoxin Problem

The mycotoxin problem received world wide attention when poultry turkeys died suddenly in Britain. Similar outbreaks occurred on ducklings also. Both were traced to a common feed derived from groundnut meal from Brazil. *Aspergillus flavus* was found in the meal and subsequently a toxin was isolated from the fungus and named **aflatoxin** (short for *A. flavus* toxin). Four related substances, aflatoxin B_1, B_2 and G_1, G_2 were found in *A. flavus* and *A. parasiticus*. They are toxic to animals and man, implicated as (to liver) hepatotoxins and are carcinogenic.

Aflatoxin production and growth are influenced by moisture content and temperature of storage. Improper storage containers leading to increase in moisture content when relative humidity is high and temperature is moderate, leads to mould growth and toxin formation. Mechanical damage during harvest contributes to contamination. Large agricultural products like rice, wheat, sorghum and corn and oil seeds, especially groundnut and soybean are affected by these moulds. If rains occur when groundnut is stacked up at harvest time, aflatoxins accumulate rendering the product unsuitable for consumption. Fortunately, aflatoxins are removed in the process of solvent extraction of oil and therefore, refined groundnut oil is safe. However, groundnut cake is used as animal feed endangering their health. Consumption of mouldy grain causes sickness and in India, outbreaks of some food poisoning cases have been reported. In Banswada district of Gujarat, consumption of mouldy corn resulted in several deaths and food samples were found to contain aflatoxin and **zearalinone**, another mycotoxin.

Besides aflatoxins, **ochratoxin** produced by *Aspergillus ochraceus* and **trichothecenes** formed by *Fusarium* species on grains have been reported to cause sickness. Historically, a sickness called **ergotism** was known in temperate countries where rye grain contaminated with sclerotia of the fungus *Claviceps purpurea*, was consumed. It produces **lysergic acid** which is a hallucinating agent, and a hormone upsetting the nervous function of the consumer.

Penicillium patulum growing on apples and other fruits produces a toxin known as **patulin**. There are also other toxic metabolites produced by fungi contaminating food materials.

Protection from mycotoxin formation is achieved by handling of produce with minimum mechanical damage, and drying the produce rapidly to bring down fungal population by rapid reduction in moisture content. There are safe moisture levels for different grains and oil seeds and whenever necessary, artificial drying by mechanical driers has to be carried out because mycotoxins are associated with widely used common foods.

11

MICROORGANISMS IN HUMAN DISEASES

The recognition that transmission of a disease from one individual to another causes epidemics was credited to Fracastonius (1546) who suggested that the spread involved passage of infective particles by direct contact of individuals. Although microorganisms were discovered by Leeuwenhoek in the next century, the role of microorganisms as infectious agents was recognized only in the second half of the 19th century, mainly through the researches of Berkeley (potato blight), Pasteur (Rabies) and Koch (anthrax and tuberculosis). Robert Koch (1843–1910), a German scientist is regarded as the father of medical microbiology. Table 21 lists the initial discoveries of human diseases caused by bacteria.

Table 21 Discoveries of Important Infectious Diseases of Humans Caused by Bacteria

Year of discovery	*Name of disease*	*Causative organism*	*Discoverer*
1877	Anthrax	*Bacillus anthracis*	Robert Koch
1879	Gonorrhoea	*Neisseria gonorrhoeae*	Neisser
1880	Typhoid	*Salmonella typhi*	Eberth
1882	Tuberculosis	*Mycobacterium tuberculosis*	Robert Koch
1883	Cholera	*Vibrio cholerae*	Robert Koch
1883	Diphtheria	*Corynebacterium diphtheriae*	Klebs
1884	Tetanus	*Clostridium tetani*	Nicolaier
1885	Diarrhoea	*Escherichia coli*	Escherich
1886	Pneumonia	*Streptococcus pneumoniae*	Fraenkel
1888	Food poisoning	*Salmonella enteritidis*	Gaertner
1894	Plague	*Yersinia pestis*	Yersin, Kitasato
1896	Botulism	*Clostridium botulinum*	Van Ermengem
1898	Dysentery	*Shigella dysenteriae*	Shiga
1900	Paratyphoid	*Salmonella paratyphii*	Schottmuller
1903	Syphilis	*Treponema pallidum*	Schaudinn and Hoffman
1906	Whooping cough (Pertussis)	*Bordetella pertussis*	Bordet and Gaegou

During the same period it became apparent that some diseases which are clearly infectious did not yield any recognizable bacteria. The demonstration by Iwanovsky (1892) that **Tobacco Mosaic** is caused by a filterable agent paved the way for recognizing viruses as agents of some diseases. Beijerinck (1898) confirmed that tobacco mosaic is caused by a virus. Loeffler and Frosch (1898) showed that **foot and mouth disease** of cattle is caused by a virus. **Smallpox, polio** and certain other diseases were later shown to be caused by viruses. Though Pasteur developed the vaccine against rabies in 1885, the nature of the virus was established much later.

Malaria was shown to be caused by a protozoan, *Plasmodium vivax* (Ross, 1902) and three other species of *Plasmodium*. **Amoebiasis** which is a widespread health problem is caused by *Entamoeba histolytica*.

NORMAL MICROFLORA OF THE HUMAN BODY

An average person has about two square metres of skin. The skin offers different micro environments, ranging from exposed dry parts to sheltered areas like the armpits. Sweat glands occur on all parts of the skin. **Sebaceous glands** occur in follicles of hair and secrete a lubricant substance. The oral cavity is a very favourable site for microorganisms as it is moist and receives fresh food regularly. The intestines also harbour a variety of bacteria, feeding on undigested food.

Microorganisms on external parts may be regular inhabitants or transient forms occurring periodically. A majority of skin microflora are **commensal** in nature, not causing any recognizable harm. However, pathogenic forms also occur. The various microorganisms compete for space and nutrients. This limits their activities. It may have a protective effect if the pathogens are inhibited due to competition. Gram-positive bacteria like *Staphylococcus* and *Micrococcus* predominate. *Candida* is the common fungus found on the skin. *Propionibacterium acnes* is the cause of acne, which develops on the face. A lipolytic yeast *Pityrosporum ovalis* occurs occasionally on the scalp. Bacteria predominate in the oral cavity. Staphylococci, lactobacilli, streptococci and also an actinomycete occur. Dental caries (tooth decay) is caused by *Streptococcus sobrinus* and *S. mutans.*

The major organs of the gastrointestinal tract (stomach, small intestine and large intestine) harbour a variety of microbial flora. Bacteria predominate and yeasts are found occasionally. The **gastro-intestinal tract** is sterile before birth but breast feeding results in lactobacilli and streptococci colonizing the tract.

The stomach is highly acidic (pH about 2.0) and bacterial population, though consisting of several genera, is very low. However, some bacterial cells escape death when enclosed in food particles and pass into the intestine. The pH gradually rises in the small intestine due to secretions of **proteolytic enzymes** and large populations of bacteria develop feeding on undigested food materials. Many of them are **facultative anaerobes**. Cell numbers of 10^5 to 10^7 per gram of food material have been recorded. The predominant genera are *Bacteroides, Lactobacilli, Coliforms* and *Streptococcus*.

Oxygen in the digestive tract is used up in the small intestine, creating anaerobic conditions in the large intestine, where water with dissolved substances is absorbed into the body. The food material gets modified by action of bacteria on **bile acids** secreted by the liver, to aid in digestion of fats. Facultative and obligate anaerobes predominate in the large intestine. A variety of metabolic reactions take place in it. Vitamins such as **thiamine, riboflavin,** and **cyanocobalamin (B12)** are produced by the intestinal flora, utilizing **lactose, cellulose** and other undigested food. *Escherichia coli* is known to produce vitamin K. Nitrogen fixation by *Klebsiella pneumoniae* under anaerobic conditions in the intestine has been demonstrated and is considered to be a source of nitrogen in certain communities whose diet consists of only carbohydrates. Anaerobic conditions in the large intestine also result in **production of gases** such as CO_2, H_2, CH_4 and H_2S resulting in odour formation in evolved gases. Air taken into the gastrointestinal tract along with food and water, which is predominantly made up of nitrogen, forms major part of the gas.

Apart from skin and the gastrointestinal tract, the **upper respiratory tract** and the **outer cavity of the ear** harbour some microorganisms. *Bacteroides, Corynebacterium, Streptococci, Branhamella* etc., have been recorded from the respiratory tract. The **urinogenital system** is generally sterile but the vaginal cavity harbours some microorganisms. Lactobacilli predominate while others are inhibited due to **lactic acid production** by those forms. Yeasts are also found and *Bacteroides* and *Corynebacteria* have also been recorded.

GENERAL ACCOUNT OF PATHOGENIC BACTERIA

Human pathogens include the widest range of morphological groups, ranging from *Streptococci* (chain of cocci), *Staphylococci* (cluster of cocci), *Diplococci* (double-celled and capsulated), *Enterobacteria* (bacilli), *Clostridia* (bacilli with endospores), *Vibrios* (bent rods), *Corynebacteria* (irregular shaped cells) and *Treponema* (spiral shaped). Actinomycetes show branched filaments. Gram-positive and Gram-negative forms occur. In terms of oxygen requirement, obligate aerobes (*Mycobacterium tuberculosis*) are fewer while many are facultative anaerobes and some are obligate anaerobes (*Clostridium*).

Temperature optimum of most pathogens is around 37°C, which is the normal human temperature. *Yersinia pestis* has a lower optimum of 28°C while *Campylobacter jejuni* grows best at 43°C. Most of them grow best in the pH range of 7.2–7.6 and growth ceases below 5.0 and above 9.0. *Vibrio cholerae* prefers alkaline conditions of pH around 8.0 to 9.0.

Different degrees of **specificity** are shown by pathogens. For instance, the anthrax bacillus is the least specific, causing a disease in animals and humans. Gonorrhoea is specific to humans. Many are confined to a particular part of the body only. Some are said to be opportunist. For instance *E. coli, Streptococcus* and *Staphylococcus* are constituents of normal microflora but become pathogenic when body defences are weakened.

Individual strains within a species show **variation** in pathogenisis.

Some strains are **virulent**, causing severe disease while others are **less virulent** giving rise to mild symptoms. Some are **avirulent**, i.e., not causing any disease. Special features of pathogenic strains enable them to overcome host defenses and cause disease. Adverse effects on hosts are either due to proliferation of the pathogen disturbing host nutrition and metabolism, or production of **toxins** which damage the host cells. The outer membrane of some Gram-negative bacteria, such as *Salmonella* acts as an **endotoxin**, harming the host.

Toxins of Pathogenic Bacteria

Many pathogens produce toxins which are harmful to the host. The toxins are grouped into **exotoxins, enterotoxins**, and **endotoxins**. The exotoxins are secreted by the organism into the surroundings. On the basis of mode of action, exotoxins are subdivided into **cytotoxins** and **neurotoxins**. Cytotoxins kill the cells by enzymatic action. They may also cause haemolysis, damaging blood cells (e.g. *Streptococcus* and *Staphylococcus* infections). The diphtheria toxin interrupts protein synthesis. The tetanus toxin is a neurotoxin which binds to nerve cells causing muscle contraction (paralysis).

Enterotoxins are also exotoxins in terms of their secretion into the surroundings. As they are produced in the gastrointestinal tract, they are treated separately. *Vibrio cholerae* produces polypeptides which cause rapid **loss of intestinal fluids**. Some species of *Salmonella* also produce enterotoxins which cause diarrhoea. However, several other species possess endotoxins which result in severe diseases (*S. typhi* causing typhoid and *S. paratyphi* causing paratyphoid). The outer **lipopolysaccharide membrane** of these Gram-negative bacteria itself is the toxin. Hence, the disease occurs only when large doses of the pathogen find their way into the gastrointestinal tract. Intestinal infections by *Escherichia coli, Shigella spp; Bacillus cereus* etc. produce diarrhoea. The exotoxins and enterotoxins produced by these pathogens are soluble proteins and **heat labile** (destroyed on heating) while the endotoxins which are on lipid membranes are **heat stable**. Table 22 lists important toxins of pathogenic bacteria.

DISEASES CAUSED BY BACTERIA

The following account discusses important diseases caused by bacteria.

Respiratory Tract Diseases

Pneumonia. Caused by *Streptococcus pneumoniae* (previously known as *Diplococcus pneumoniae*). The presence of a capsule makes it resistant to phagocytosis. It affects the respiratory tract and lungs causing inflammation. The disease usually occurs as a secondary infection when host defences are weakened by other infections.

Pharyngitis (Sore throat) and Rheumatic fever. The causative

Table 22 Toxins of Pathogenic Bacteria

Organism	*Disease*	*Toxin*	*Chemical nature and properties*
		EXOTOXINS	
Corynebacterium	Diphtheria	Diphtheria toxin	Protein, Mol. Wt. 65000. Blocks amino acid transfer to peptide. Lysogenic phage codes for the toxin.
Clostridium	Lockjaw	A neurotoxin	Protein, Mol. Wt. 67000. Binds to lipid in nerve cells causing muscle contraction (paralysis).
Streptococcus pyogenes	Boils etc., tonsilitis, scarlet fever-(rash)	Streptolysin-*O* Streptolysin-*S*, Erythrogenic toxin	Lysis of membranes and haemolysis -do- -do- (all are proteins)
		ENTEROTOXINS	
Vibrio cholerae	Cholera	An enterotoxin and mucilage enzymes (controlled by chromosomal genes)	A polypeptide with A_1, A_2 and B chains (controlled). Intestinal fluid loss.
Staphylococcus aureus	Boils etc., diarrhoea	Several toxins	Protein, Mol. Wt. around 27800 haemolysis, death of leucocytes, diarrhoea.
Salmonella spp.	Intestinal infections	Similar to cholera toxin	Diarrhoea
		ENDOTOXINS	
Salmonella spp. (S. *typhi*, *S. paratyphi*, *S. enteritidis etc.)*	Typhoid, paratyphoid, intestinal disorders etc.	Lipopolysaccharide membrane of bacteria.	Large doses of the pathogens are required to produce disease.
Escherichia coli	Intestinal infections	-do-	-do-
Shigella spp.	-do-	-do-	-do-
Bacillus cereus	-do-	-do-	-do-

agent in both these conditions is *Streptococcus pyogenes.* Sore throats are also caused by other agents such as viruses. Hence, proper diagnosis is necessary for treatment. *Streptococcus pyogenes* produces proteinaceous toxins known as **streptolysins,** which affect mucous membranes and also cause **haemolysis**. Toxins are produced only in strains which carry a lysogenic bacteriophage in them.

Streptococcus pyogenes also causes **rheumatic fever**. Such strains do not have the lysogenic phage in them. Their cell surface antigens resemble those of host immune chemicals. Antibodies produced by the immune system themselves act on the body. Hence, the disease is known as **autoimmune disease**. Tissues of the joints and the heart are damaged by these antibodies. Kidney damage also occurs. Rheumatic patients should, therefore, keep a watch on heart and kidney condition in later stages.

The spread of these pathogens is through contact with people who are carriers, and through inhalation of dust etc. carrying the germs. Both the forms are highly sensitive to penicillin and treatment is simple if diagnosed early.

Diphtheria. *Corynebacterium diphtheriae* is the causative agent. It lodges in the throat and tonsils causing inflammation. A pseudomembrane is formed by damaged host cells and the bacteria. This may block the passage of air. Death results both from damage to cells and suffocation. Strains lysogenized by bacteriophage β, produce a powerful **exotoxin.** A non-toxic **toxoid** is prepared from this toxin and used as a vaccine for prevention of the disease. This toxoid is usually administered as a **combined DPT vaccine** (toxoids of diphtheria and tetanus and killed cells of pertussis).

Tuberculosis. This highly destructive and widespread disease is caused by the actinomycete, *Mycobacterium tuberculosis.* This organism was first isolated by Robert Koch in 1882. Tuberculosis is the cause of many deaths worldwide. The bacterium is acid-fast, non-motile and aerobic. Five types are known. Human, bovine, avian, murine and the one on cold-blooded animals. The cells are slender bacilli occurring singly or in clumps with cells attached at an angle with each other, forming rope-like colonies in culture. Its nutritional requirements are complex and it is cultured on **Lowenstein-Jensen medium** containing egg yolk, glycerol, asparagine and malachite green.

Tuberculosis is a **slowly progressive disease** which often affects the lungs, while other organs are also sometimes affected. Severity of infection is a result of interaction between the strain and resistance of the infected individual. Spread of the disease is chronic, involving destruction of lung tissue, followed by partial healing. This is revealed in X-ray photographs. Malnutrition, stress, and hormonal imbalance are cited as factors predisposing individuals to attack. A diagnostic test, known as the **tuberculin test** is performed and individuals showing a positive hypersensitive reaction on the skin are interpreted as having been infected by the bacterium, although they might not have developed the disease.

Besides infection of the lungs, *M. tuberculosis* also causes other types of diseases. Infection of brain cells results in **tuberculous meningitis. Bone and joint tuberculosis** affects many different sites in the body. **Miliary tuberculosis** is a blood disease affecting different parts of the body due to the pathogen spreading through blood.

Inhalation of air carrying dust particles or spray from the sputum of infected individuals is the usual source of infection. Milk carrying bacilli from bovine infection is another source. Observing primary health measures, destruction of cattle suffering from bovine tuberculosis and avoiding overcrowding are advocated as preventive measures. **Chemotherapy** is now fairly effective, especially after the discovery of antibiotics active against mycobacteria. Streptomycin and Rifamycin are widely used. Other drugs like isoniazid and para aminosalycilic acid are employed in combination therapy.

Whooping cough (Pertussis). This is a severe disease affecting children, especially under the age of one year. It is caused by a Gram-negative, aerobic, coccobacillus (short rods) called *Bordetella pertussis.* It attaches to the cells of the upper respiratory tract. It produces an exotoxin which damages host cells. The symptom is a recurring spasmodic cough, hence the name whooping cough. Killed cells of the pathogen are employed for vaccination (a component of DPT vaccine). The disease has been brought under control by universal vaccination with DPT vaccine.

Infections of the Gastro-Intestinal Tract

The normal intestinal flora consist of Gram-negative, aerobic or anaerobic bacteria which ferment glucose and are **catalase-positive.** Most of them are commensals, and not harmful, but some of them are potential pathogens. Active pathogens also find their way into the tract through contaminated food and water. They are collectively known as **enteric bacteria**. The important genera which are regular pathogens or opportunists are *Escherichia, Klebsiella, Proteus* and *Vibrio.*

Enteric bacteria show complex antigenic properties and many serotypes are known. The antigens are grouped into **H antigens** (flagellar proteins), **'O' antigens** (cell wall constituents) and **'K' antigens** (capsular). They also possess **colicins**, which are antigenic proteins and inhibitory to related bacteria in mixed infections. The lipopolysaccharide outer membrane of some forms is toxic (endotoxins), while some others form exotoxins.

Salmonella. The genus has several species and within each species there are many antigenic types. About 1800 serotypes are known. *Salmonella typhi* causes **typhoid fever**. A progressively mounting fever, headache and weakness are the symptoms. After some time **rose-spots** appear on the abdomen and diarrhoea occurs. The organism spreads from the intestine to the blood and other organs. In severe cases, intestinal perforation and haemorrhage result and may be fatal. It is a predominantly **water borne**

disease and large doses of the pathogen are needed for severe infection. The antigens stimulate immunity in the exposed individual, but control is difficult because of the large variation in the antigens. Culturing of blood is a diagnostic method useful in the first week of the disease. Avoiding contaminated water is the preventive method. Therapy is through the antibiotics chloramphenicol and tetracyclines.

Paratyphoid is caused by *S. paratyphi, S. schotimulleri* and *S. hirshfeldii.* The disease resembles typhoid fever but the symptoms are milder and duration in the body is shorter. Fever and abscess on infected site causes *septicaemia.*

Salmonella **food poisoning** and diarrhoea are caused by several species of *Salmonella,* such as *S. enteritidis,* and *S. newportii.*

Shigella. A frequent cause of diarrhoea is contamination by species of *Shigella.* They are confined to the intestinal tract. *S. dysenteriae* is the type species. Others are *S. flexneri, S. boydii* and *S. sonnei.* They are Gram-negative, non-lactose fermenting and non-gas forming facultative anaerobes. About 40 serotypes based on 'O' antigens and lipopoly-saccharides of *S. dysenteriae* are known. They cause necrosis of the mucous membrane resulting in superficial ulceration. Symptoms are, abdominal pain, fever, and watery diarrhoea. Symptoms subside in 2–3 days in healthy individuals but in children and old people infusion of electrolytes is necessary to overcome dehydration. An **endotoxin** and an **exotoxin** are produced by the bacterium. Control is by administration of antibiotics (chloramphenicol, tetracycline and ampicillin). *Shigella spp.* are transmitted through contamination of food and water by faecal matter.

Yersinia. *Yersinia pestis* is the cause of plague. It is a Gram-negative rod, showing bipolar staining (stain confined to the poles). It is a facultative anaerobe with an optimum at 28°C. The lipopolysaccharide membrane has toxic effect **(endotoxin)**. A surface protein has **anti-phagocytic** effect. Rodents act as reservoirs of the bacterium and transmission is through fleas, which suck their blood. Based on symptoms, plague is described as **bubonic** in which lymph glands are enlarged (bubos), **pneumonic** which develops when the bacterium establishes a focus in the lungs, and **septicaemic** which is characterised by hypotension, vomiting and diarrhoea. Plague, unless treated promptly, results in cardiac arrest and death. Streptomycin and tetracyclines are effective. Rodent control is a preventive measure.

Vibrio. The name vibrio is derived from the vibrating movement of the cells in water. The cells are **curved rods** (comma shaped) with **polar flagella**. They are facultative anaerobes, Gram-negative and carbohydrate-fermenting. *Vibrio cholerae* is the cause of cholera. Other species include *V. parahaemolyticus* and *V. anguillarum.* They resemble enteric bacteria in some respects and are sometimes included under *Enterobacteriaceae,* while others retain them in the family *Vibrionaceae.*

V. cholorae is a typical curved rod, 1.5 × 0.5 μm. Occasionally the cells occur in pairs, forming S-shaped structures. The optimum temperature is 37°C and the optimum pH is 8.2. It produces M and O antigens. The latter are distinctive to *V. cholorae* and more than 100 strains in which O antigens are present are known to cause cholera. The pathogen produces a **proteinaceous enterotoxin** (Mol. wt. 80000). Cholera results when large doses of the bacterium are ingested (10^8 to 10^{10} cells) mainly through faecal contamination of water.

The disease is characterised by sudden onset of nausea and vomiting, profuse diarrhoea and abdominal pain. The stools have characteristic **rice-water** appearance (rice stools). The bacterium is confined to the intestinal tract. It stimulates hypersecretion of water and chlorides. As a result, **loss of electrolytes** and **dehydration** occur causing death. Because of the spread of the disease is rapid, widespread epidemics have occurred in many countries. The best method of management is prevention through proper sanitation and consumption of clean water and food. Vaccination with **cholera toxoid** is given as a prophylactic measure whenever the disease makes its appearance. Replacement of water and electrolytes is carried out on affected individuals.

Escherichia. *Escherichia coli* is a common inhabitant of the intestinal tract but occasionally becomes pathogenic, especially in the **urinary tract. Coliform test** is carried out on water as a measure of bacterial contamination because it is a regular intestinal inhabitant. Pathogenic strains form O antigens and K antigens. They cause diarrhoea especially in children.

Tetanus. *Clostridium* is an obligate anaerobe, widespread in nature. Some species are pathogenic to humans and animals. It is a spore-forming genus and shows characteristic drumstick appearance, which can be seen in culture when endospores are formed terminally in cells. *Clostridium tetani* causes **tetanus**, *C. perfringens* produces **gas gangrene** and *C. botulinum*, **food poisoning**.

Infection of *C. tetani* is localized and results from contamination of wounds with soil or other contaminated materials. As it is an anaerobe, it develops in deep wounds or other concealed areas, in surgical operations etc. The effect is mainly due to the toxin, **tetanospasmin** produced by the bacterium. Muscular spasms occur due to inhibition of muscle movement. Muscles in the jaw are affected and gradually other muscles are affected resulting in **lock-jaw**, pain and blocking of respiration. The mortality rate is nearly 50%.

Immunization with **tetanus antitoxin** is practised whenever infection is suspected, and as a prophylactic measure in operations. Antibiotics such as penicillin are effective against the bacterium.

Diseases of the Urinogenital Tract

These are generally referred to as **sexually transmitted diseases (STD)**.

They present a particularly difficult problem for public health people because of reluctance of sufferers to give information.

Gonorrhoea. This is caused by *Neisseria gonorrhoeae.* Both males and females are affected but painful symptoms appear in males due to infection of the urethral canal. Vaginal infections cause mild symptoms. Newborn babies may be affected. Penicillin ointment is applied to control infection in infants. The disease can be effectively treated with penicillin or other antibiotics.

Syphilis. Syphilis is caused by the spirochaete, *Treponema pallidum,* an obligate anaerobe. Infection is through tiny breaks in the skin. It forms lesions at initial sites of infection. Sometimes the bacterium spreads to other parts of the body such as mucous membranes, eyes, joints and the central nervous system. Hypersensitive reaction in the form of skin rash occurs. The disease may disappear but in some individuals, the bacterium persists and in pregnant women the foetus may be infected. Treatment with penicillin is highly effective in the early stage of the disease but longer treatment is necessary in the later phases.

AIDS. In recent times, AIDS has become the most serious sexually transmitted disease. We have already discussed this in the general account on viruses.

DISEASES CAUSED BY FUNGI

Fungi cause skin infections which are grouped into **subcutaneous mycoses, superficial mycoses** and **deep mycoses,** on the basis of location and spread of the fungus. Subcutaneous infections are caused by entry into the feet and into wounds by soil-borne, saprophytic fungi such as *Cladosporium, Phialophora* and *Sporothrix.* **Madura foot** is a serious condition caused by *Madurella mycetoma.* Swelling of the foot and degeneration of tissue occurs resulting in deformation. These pathogens may be described as **opportunistic pathogens.**

Superficial mycoses are caused by fungi which colonize superficial keratinized tissue such as skin, hair and nails. The main pathogens are *Trichophyton, Microsporum* and *Epidermophyton.* They are conidial stages of ascomycetes. **Ringworm** is a condition caused by these fungi. *Trichophyton* attacks various parts of the body and the condition is variously described. **Athlete's foot** (Tinea paedis) occurs in toe-webs, sometimes in combination with *Epidermophyton.* Itching ensues and small vesicles develop which discharge thin fluid. Peeling of the skin occurs and the organ gets infected with bacteria. Ringworm of the non-hairy skin (Tinea carporis) forms **circular lesions** with a scaly centre surrounded by red coloured and advancing border, often containing vesicles. Ringworm of the scalp (Tinea capitis) is caused by *Microsporum.* Small bald patches occur on the scalp. It is common in children. Infection of nails (Tinea unguinum) is caused by several species of *Trichophyton*. Nails become thickened and discoloured.

Dermatophytic fungi are highly resistant to antibiotics and treatment is difficult. Treatment consists of cleaning and removal of dead tissues and application of antifungal chemicals like salicylic acid (3%) or benzoic acid (5%) in a creamy base.

Histoplasmosis is a condition of deep mycosis, caused by *Histoplasma capsulatum*. Other deep mycoses are **blastomycosis** (by *Blastomyces dermatidis*) and **coccidiomycosis** *by Coccidioides immilis*. **Histoplasmosis** affects endothelial tissues, especially in the respiratory tract and lungs. Spread is through inhalation of conidia of the fungus. Blastomycosis results in ulceration of internal tissues and bones and also skin lesions. Its spread is also through inhalation of conidia. The dermatophytes are controlled with the antibiotics **amphotericin B** and **griseofulvin** which distort mycelial growth, thus interfering with their growth.

Candidiasis. *Candida albicans* is a yeast placed in deuteromycetes (no ascus stage is found). It occurs as a saprophyte in many parts of the body and behaves as an opportunistic pathogen. It is a budding yeast with spherical or oval cells (3–5 μm × 5–10 μm), sometimes forming pseudomycelia by chain-like attachment of cells. Depending on the organ of attack, the disease is known as **bronchial** or **mucocutaneous**, or **vulvo vaginitis** etc. As *Candida* is also highly resistant to antibiotics, the best method of prevention is by allowing natural microflora of the body to grow so that the population of *Candida* is kept low. Superficial infections can be treated with the **polyene antibiotics** nystatin and amphotericin B.

DISEASES CAUSED BY PROTOZOA

Two diseases which are widespread in tropical countries i.e. **malaria** and **amoebic dysentery** are caused by protozoan parasites. **Trypanosomiasis** (sleeping sickness) is another protozoan disease prevalent in Africa. Several other less widespread diseases are also caused by protozoa.

Malaria. Malaria is caused by species of *Plasmodium*, belonging to the class **sporozoa**. Four species are involved, of which *P. vivax* is the most important. Its life cycle alternates between humans and the mosquito vector, the latter being a species of *Anopheles*. As mosquitoes are prevalent in the tropics, the disease occurs in many tropical countries.

The parasite has a complex life cycle. As the mosquito sucks blood from an infected individual, it picks up gamete cells present therein. These cells develop into male and female gametes. The zygote that results from fertilization enlarges and forms a number of **sporozoites**. They are released and reach the salivary glands of the mosquito. When it bites a healthy individual, some of them are transmitted. The sporozoites reach the liver and form minute units known as **merozoites**. These enter the erythrocytes (red blood cells) and reproduce, forming gametic cells. The gametic cells are picked up by the mosquito, and the life cycle repeats. Figure 65 illustrates the life cycle of *P. vivax*.

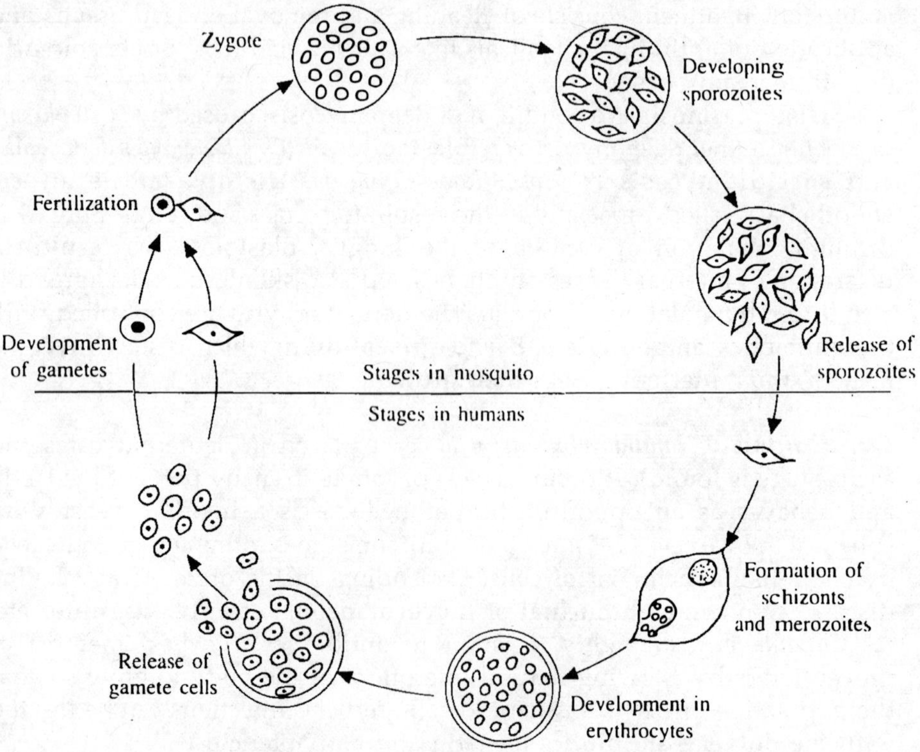

Fig. 65 Life cycle of the malarial parasite (*Plasmodium vivax*).

Symptoms develop in about 48 hours after infection (as merozoites are released in a 48 hour cycle), and are characterised by fever and chills. Anaemia results due to loss of red blood cells. **Chloroquine** is used for treatment but the parasite is not completely eliminated because chloroquine acts only on the erythrocyte stage. Other drugs have been developed recently but none of them can completely cure the infection. Worldwide mosquito eradication programmes have been carried out with the help of the World Health Organization (WHO). DDT was used as spray on stagnant bodies of water where the mosquito breeds. It was effective for some years but eventually failed because the mosquitoes developed resistance to it. Other chemicals are now in use. However, malaria continues to be a problem in tropical countries.

Amoebic dysentery. *Entamoeba histolytica* is the cause of amoebic dysentery. It is an **intestinal parasite** infiltrating the tissues, particularly of the large intestine, and causing ulceration and acute diarrhoea. The disease is a chronic problem occurring for years together in infected individuals due to the parasite's ability to form **cysts** These are highly resistant and in a dormant state. Occasionally, the cysts germinate and the resulting active forms cause diarrhoea. Faecal examination reveals the presence of cysts and active amoeboid form. When infection is severe, abdominal pain is felt. Mucus and blood are discharged with faeces and this stage is referred to as dysentery.

Transmission of the parasite is through contaminated water and occasionally, food. Control is difficult because the cysts are highly resistant to drugs. When dysentery occurs, active forms are controlled by **Metronidazole**, which is generally administered in combination with other drugs.

Skin Infections

Staphylococcus aureus, Streptococcus pyogenes and *Pseudomonas aeruginosa* are frequently occurring pathogens associated with boils, blisters, carbuncles, wounds and burns. They are opportunistic organisms forming a part of the normal skin flora but becoming pathogenic when the body is subjected to some adverse condition. They are **pyogenic** (pus-forming). Presence of blue pus is indicative of infection by *P. aeruginosa*. All the three are associated with burns and are frequent cause of death of individuals receiving burn injuries. *Staphylococcus aureus* is a common cause of infections in hospitals.

EPIDEMIOLOGY AND CONTROL OF DISEASES

A disease is considered to be an epidemic when a large number of individuals in a community are infected. On the basis of the area of occurrence, epidemics are said to be at three levels, namely **endemic** diseases, which are confined to a restricted area but are regularly occurring, **epidemic,** when a large number of individuals are suddenly affected, and **pandemic,** when an epidemic is of widespread occurrence, often extending across countries. A disease that occurs occasionally and on fewer individuals is said to be a **sporadic** disease. The study of conditions leading to, and methods of, containing the spread of diseases is known as **epidemiology**.

Several factors are involved in the development of epidemics. Infectivity and reproductive capacity of the pathogen are important. Virulence of the pathogen, severity of the disease, nature of transmission, agents of transmission and prevalence of enviromental conditions favourable to disease spread are important. A pathogen which has a short incubation time in the host and a high rate of production of cells or spores, and an agent (wind, water, food, insect vector etc.) that transmits the pathogen rapidly leads to the development of epidemics. The **primary source** of the pathogen is a **reservoir** where it is normally found. Identification of such a primary source of the pathogen helps to eradicate it so that an epidemic can be prevented. When environmental conditions are favourable and infection occurs, the affected individual is regarded as the **secondary source** of infection. Such individuals are identified by routine surveys of populations. The role of **carriers** (those not showing disease symptoms) has been found to be important in diseases such as tuberculosis and typhoid. Besides humans, animals also transmit diseases (tuberculosis, anthrax etc.) and may be primary sources of infection. Some may have more complex mechanisms. For instance, the malarial parasite not only spreads rapidly but also multiplies in the alternate host, mosquito, thus enhancing the rate of spread.

The pathogens are transmitted by air currents, water, contact and through vectors. Secretions such as saliva and mucus of infected individuals are important in the spread of tuberculosis, pneumonia and in many viral diseases (chicken pox, measles, influenza). Physical contact between healthy and infected individuals is also an important factor in some diseases (syphilis, gonorrhoea, AIDS, skin lesions etc.) Contaminated water is an important source of many pathogens such as enteric bacteria, *Entamoeba*, polio and hepatitis viruses etc. The cholera organism can survive for long periods in bodies of water and spreads rapidly under favourable conditions. Increase in mosquito populations in stagnant water is a constant source of the malarial parasite. Sometimes inanimate objects such as surgical tools in hospitals, agricultural implements, eating utensils etc. carry infectious germs.

Host resistance is of significance in the development of epidemics. Initial infection results in explosive occurrence of disease but gradually host resistance develops and a balance gets established between the pathogen and the community so that the disease does not reach epidemic proportions. The general level of resistance or immunity (freedom from infection) to disease in a community is known as **herd immunity**. Analysis of this feature is important in explaining host resistance. Herd immunity was estimated that 90–95% of individuals are resistant to smallpox, while herd immunity to polio is estimated at about 70%. Analysis of seasonal occurrence and geographical distribution serve as useful guides in predicting epidemics. Diseases such as pneumonia, influenza and chicken pox are more common in cooler months while diarrhoeas are more prevalent in summer.

The control of epidemics is based on the analysis of the disease cycle with the objective of recognizing the weakest link in the spread of the pathogen and taking suitable action. The control measures may involve direct action on the primary source of infection, on transmission of the pathogen, or by raising the immunity level of the community. Elimination of rabies-carrying animals (mostly dogs and foxes) has been in practice. In the case of humans acting as reservoirs (diseased individuals or carriers), therapy with drugs is the only practical way of reducing the primary source (tuberculosis, STD pathogens, wound pathogens etc.). Quarantine regulations have been established to prevent the movement of infected individuals and vectors from one region to another. Measures of sanitation and water purification are the common public health activities aimed at controlling the pathogens at the transmission stage. **Pasteurization** of milk is aimed at achieving this objective. Destruction of vectors (mosquitos, rodents etc.) is an important practice. For most of the highly infectious diseases, control of epidemics is sought to be achieved by raising the immunity level of communities. Spectacular success has been achieved in the control of diseases for which effective vaccines are available. Smallpox has been completely eliminated by universal vaccination. At present, efforts are being made to eliminate polio completely by the year 2000, by vaccinating all children. Permanent immunity occurs in these vaccinations. DPT vaccination is protecting children from the three diseases, diphtheria, pertussis and tetanus. Where immunity lasts only for a limited period, vaccination of entire communities is carried out whenever conditions for spread of disease are predicted to be favourable. For instance, whenever cholera cases are detected, vaccination is carried out extensively to prevent the epidemic. This is because cholera toxoid gives immunity only for about six months.

CHEMOTHERAPY

The control of infectious agents (bacteria, fungi, protozoa) except viruses is carried out to a large extent by the application of drugs which act on them. The concept of chemotherapy was enunciated by Paul Erlich (1907) who had proposed that infections can be controlled by substances with selective action, which inhibit bacteria but do not have adverse effects on higher organisms. No drug fully satisfies this condition but a host of drugs have now been developed to control pathogens.

A chemotherapeutic substance should, ideally be inhibitory to the microorganism in very low doses, compatible with mammalian tissues (of humans and animals) producing least side effects, and when administered orally (tablets, capsules, liquids) or parentarally (injections, drips), should reach the sites (receptor sites) where the microorganism is lodged, and leave the system slowly (24–48 hrs) through the excretory systems. Also, it should be economical to produce. Because of these limitations, only a small proportion of the thousands of antimicrobial substances discovered have come into clinical use. Such an ideal substance is (figuratively) referred to as **Magic Bullet**. However, no perfect magic bullet has so far been discovered but penicillin comes closest to it.

Substances produced by microorganisms and inhibitory to other microorganism were termed as **antibiotics**. Nowadays substances extracted from plants or animals are also called antibiotics. However, antibiotics from microorganisms alone have come into clinical use. **Antiseptics** are substances that are general poisons of protoplasm, inhibiting living cells without the property of selective toxicity. They are mainly useful in **topical application** (use on external organs) as ointments, sprays, dusts etc. They are also employed to sterilize instruments and hospital rooms. When they are used for the latter purpose, they are called **disinfectants**.

Antiseptics and Disinfectants

The following groups with examples are listed:

- Phenols and related compounds (phenol, cresol, hexachlorophene);
- Halogen-containing compounds (chlorine, hypochlorites, iodine);

- Acids (benzoic acid, boric acid);
- Alkalies (sodium or potassium hydroxide);
- Alcohols (ethanol, isopropanol);
- Aldehydes (formalin, glutaraldehyde);
- Surfactants (soaps, polysorbates);
- Oxidizing agents (hydrogen peroxide, potassium permanganate);
- Dyes (crystal violet, methylene blue, proflavin);
- Fumigants (ethylene oxide) and
- Heavy metals (silver nitrate, mercuric chloride).

The mechanisms of action of antiseptics and disinfectants include **coagulation of proteins, altering the properties of bacterial cell walls and inactivation of enzymes by binding to them.**

Antimicrobial Substances

The chemical nature of **antibiotics** is diverse (some important antibiotics were considered in the chapter on industrial microbiology). The clinical use of antibiotics as well as that of sulphonamides and some other synthetic chemicals has revolutionized treatment of people and vastly improved their health all over the world. One major problem is the development of drug resistance in microorganisms. Hence, continuous search for newer substances or modification of existing ones is going on. For instance, **semisynthetic penicillins** (ampicillin, cloxacillin), and **cephalosporins** (cephalaxin, cephaloridin) are now in extensive use.

The discovery of **sulphonamides** by Domagk (1935) and their quick adoption for clinical use has greatly facilitated chemotherapy. Sulphonamides act by substituting for **PABA (para aminobenzoic acid)** which is required for folic acid synthesis in bacteria. They are mainly useful in controlling intestinal disorders. Sulphadiazine and sulphamethoxazole are commonly used.

Trimethoprim, a pyrimidine derivative, in combination with sulphonamides has proved to be very effective against a wide spectrum of microorganisms and is widely used now.

Nitrofurans are active against Gram-positive and Gram-negative bacteria and yeasts. Nitrofurazone is applied topically as an ointment on wounds and burns. Nitrofuroxine is employed to control vaginal infections by the yeast *Candida albicans*.

Mode of Action. The mode of action of different antibiotics brings out their effects on microorganisms. Penicillins and cephalosporins (β-lactam antibiotics) inhibit cell wall synthesis; amphotericin B, polymyxin etc. disrupt membrane integrity, causing loss of electrolytes and macromolecules; aminoglycosides (streptomycin, gentamicin etc.) attach to ribosomes, and tetracyclines attach to tRNA, thus inhibiting protein synthesis; while actinomycin and rifamycin inhibit nucleic acid synthesis by binding to DNA-dependent RNA polymerase.

12

BASIC METHODS IN MICROBIOLOGY

The basic techniques in the study of microorganisms are **microscopy** and **pure culturing**. Detailed observation of microorganisms is possible only at high magnifications of a compound microscope. The development of the electron microscope has further increased the scope of observation of the fine structure of microorganisms by greatly increasing the levels of magnification. Observation of microorganisms in the tissues of higher organisms on which they occur is largely due to many refinements that have been developed in microscopy. Detailed observation of microorganisms is facilitated by separating them from other organisms and from extraneous matter. This is achieved by methods of pure culturing which involves preparation of nutrient media and materials for culturing the microorganism, its isolation and growth on a suitable medium, and when a mixture of organisms occur, separation of the organism to bring it into pure culture, so that others are excluded. Studies on microbial metabolism, pathogenic behaviour, genetics, relation to environment, industry and the control of microorganisms have advanced through the development of biophysical and biochemical techniques, statistical methods, and mathematical modelling, fermentation and other bioengineering techniques, sophisticated instrumentation, automation, and computerization. Hence, it is not possible to describe the full range of techniques and only the basic techniques of microscopy and culturing are presented in this chapter.

MICROSCOPIC METHODS

Microscopes are optical devices based on visible light, ultraviolet radiation or the use of electromagnetic lenses. Direct observation of microscopic objects is possible with **light microscopes** based on visible light, and **fluorescence microscopes** in which UV light passes over a fluorescing substance to highlight the object under observation, avoiding direct impact of harmful UV rays on the eyes. The highest magnification that can be achieved with light microscopes, (due to restriction of wavelength range of visible light) is about 1000 μm and the smallest object that is visible is 0.1 μm. Slightly higher magnifications are possible with fluorescent microscopes because UV radiation has a shorter wavelength range than visible light. Electrons released from a source of radiation have much shorter wavelengths, hence, much higher magnifications of up to 3,00,000 were achieved with electron microscopes. However, due to the harmful nature of electrons, direct observation of living specimens is not possible. The specimens are dried and coated with an electron-dense substance before being photographed in an electron microscope.

Light microscope. A microscope consisting of two sets of lenses is known as a **compound microscope**. It consists of a stand on which a tube is mounted in such a way that it can be moved vertically by a 'rack and pinion mechanism. A flat stage with a central opening is fixed to the stand. Two sets of magnifying lenses, comprising the eye piece, are inserted at the upper end of the tube, and objective lenses arranged on a rotating disc are attached at the lower end. A condenser lens and mirror are provided below the stage. The condenser can be vertically moved by operating a lever. The tube is moved by two levers, one for coarse adjustment and the other for fine adjustment. The specimens for observation are mounted on a glass slide, which is then placed on the stage. From a light source (sunlight or an illuminator), light is deflected vertically to reach the specimen through the condenser to pass through the objectives and eyepieces, where it gives a magnified view of the specimen. It can be directly observed or photographed by attaching a camera device to the tube. Figure 66(a) illustrates the path of light in a compound microscope.

Microbial cells, spores, mycelia etc., can be suspended in a drop of water or a nutrient liquid and observed in the living state. As the living structures are generally colourless, they are treated with suitable **stains** (coloured dyes) to show the overall structure, and to recognize internal structures. A number of stains are available and those commonly employed for routine examination are **methylene blue** and **crystal violet** for bacteria and **acid fuchsin** and **trypan blue** for fungi. Permanent preparation of specimens for storage is obtained by mounting the specimen in a non-evaporating mounting medium such as Canada balsam or epoxy resin. Special devices have been developed to enhance the contrast between the specimen and background in the mount. **Phase contrast microscopy** uses special condenser rings and special objective lenses to bring out external and internal structures in unstained living specimens. Nuclei, inclusion bodies, flagella etc. of eukaryotic cells can be clearly observed in phase contrast. Another system, known as **dark field microscopy** illuminates the specimen so that it appears very bright in a dark background. This method is particularly useful for counting bacterial cells.

Counting devices such as **haemocytometer** or **Neubar's chamber** aid in quantitative estimation of cells in liquid. The microscope is calibrated with a micrometer. An **eyepiece micrometer**, a glass disc with fine grid markings is placed inside the eyepiece. A **stage micrometer** having parallel lines marked on it is placed on the stage and focused. The length (or area) between adjacent markings as they appear in the eyepiece is determined by comparing with the marking as they appear on the stage micrometer. It

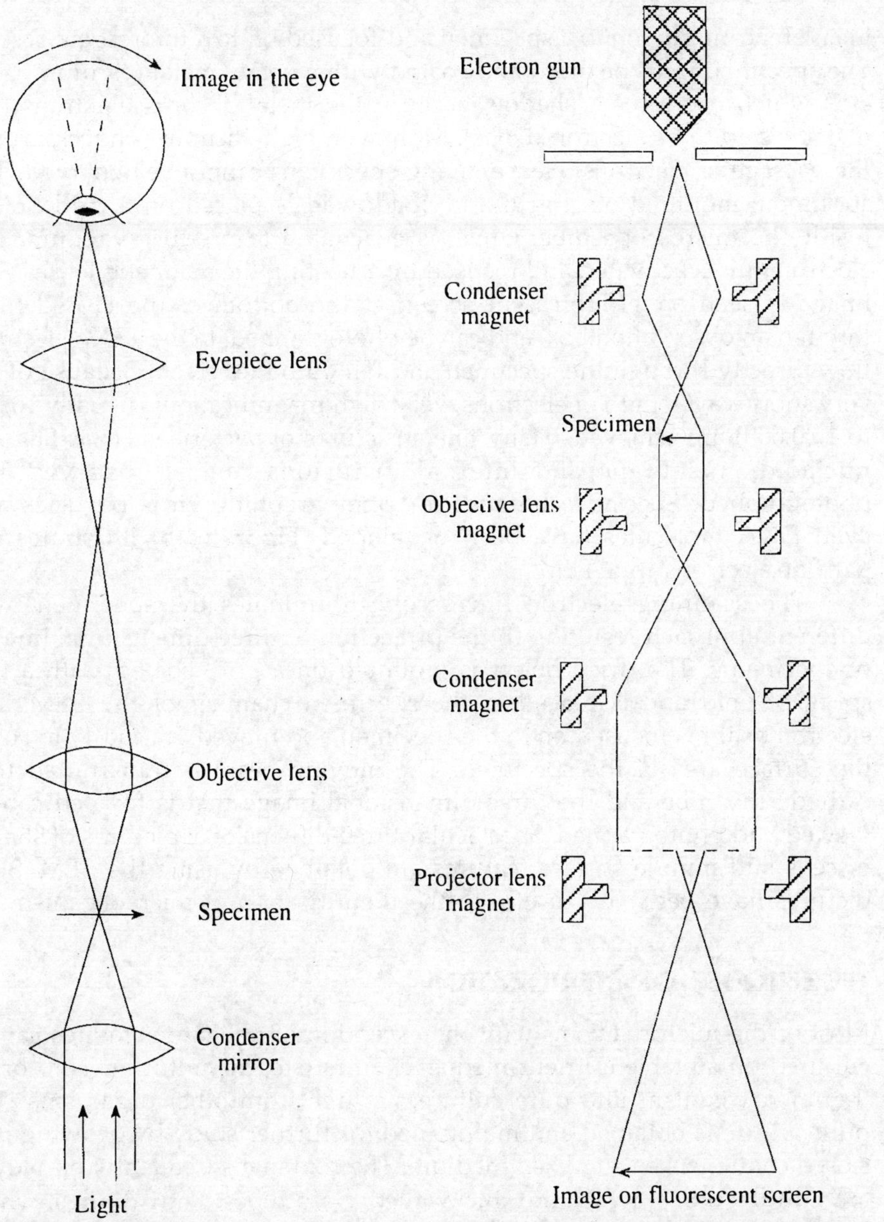

(a) Light microscope

(b) Transmission electron microscope

Fig. 66 Magnification of specimen image through light microscope and transmission electron microscope.

varies with each magnification, when different objective lenses are used for focusing. Thus the number and size of cells etc. can be determined.

Electron microscopes. There are two major types of electron microscopes, the **transmission electron microscope (TEM)** and the **scanning electron microscope (SEM)**. They use magnetic lenses to beam electrons generated in an **electron gun** on to a specimen and focused on to a fluorescent screen. The specimen must be dried and coated with a heavy metal (osmium, gold etc.) which produces a shadow because the metal absorbs electrons (the metal is said to be electron-dense). Moreover, high vacuum is necessary for the passage of electrons. Hence, living specimens cannot be observed. The specimen, mounted on a **grid** and **shadowed** is placed on a grid holder inside the microscope tube, which is evacuated to produce vacuum. The electron gun is activated and focused by adjusting the magnetic lenses. The image is seen on a fluorescent screen at the bottom of the closed tube, through an observation box, and can be photographed. In the TEM, electrons flow directly through the specimen and fall onto the screen. Because of the very short wavelength of electrons, very high magnifications (usually 10,000 to 1,00,000) are achieved. Many fine structures of bacteria such as flagella, nucleoid, membrane and internal inclusions can be observed and photographed. Special techniques for photographing virus particles and even DNA molecules have been developed. Figure 66(b) illustrates the path of electrons in a TEM.

The scanning electron microscope illuminates the specimen by a different approach resulting in the projection of three-dimensional images of specimens. The dried object is mounted on a grid, coated with a fine spray of gold and is placed in the specimen chamber of the SEM. The electron beam is focused on to the specimen and moved gradually to cover the surface area of the specimen. The moving beam is transmitted to a cathode ray tube and the three-dimensional image that is formed can be viewed and photographed. Spectacular three-dimensional pictures of shapes of cells and minute surface features are obtained by using the SEM. Such pictures have been very useful in the identification of microorganisms.

TECHNIQUES OF STERILIZATION

Most of the microrganisms with the exception of obligate parasites can be cultured on suitable nutrient media. The first step in culturing is to bring the microorganism into pure culture, i.e. free from other organisms. The pure culture is obtained and maintained for further study by growing it in closed containers on sterilized medium. The containers frequently employed are **culture tubes**, which are thickwalled glass tubes, **Petri dishes**, which are flat, circular dishes with a lid; **Erlenmeyer flasks**, which are conical flasks with a narrow mouth and flat bottom. For the transfer of microorganisms from liquids, pipettes are used and for solid surfaces inoculation needles are used. The inoculation needles are made with platinum or nichrome wire, fixed to a glass or metal rod. The tip of the needle is made either flat or bent into a loop. Either type may be used as required. The culture tubes or conical flasks are closed with a non-absorbent cotton plug or a metal lid in order to prevent contamination from other organisms. The lid of the Petri dish serves the same purpose.

The glassware is sterilized by washing, drying and heating in a hot air oven at 160° C for about two hours. This is known as **dry sterilization**. In recent times ready-to-use plastic ware presterilized by fumigation (with ethylene oxide) has come into use.

The culture media are **wet (moist) sterilized** because they are either liquids or semi-solid gels containing water. An **autoclave**, which is a boiler with an inner chamber and a tight-fitting lid is employed for this. Water is taken at the bottom between the inner chamber and outer jacket, and arrangement is made to heat the water to produce steam. The autoclave is provided with a safety valve and a pressure gauge. The principle of moist sterilization is to produce steam which raises the pressure in the inner chamber. The steam temperature goes up with increasing pressure. A constant pressure is maintained by adjusting the heating. The most frequently adopted wet sterilization temperature is 121°C, which is achieved by maintaining a pressure of 15 lbs/sq. inch (1.05 kg/sq.cm.). 15 minutes at this temperature are sufficient for sterilization.

Moist sterilization by filtration is also carried out by passing the medium through a fine filtering pad. *Cellulose acetate* filters which hold back bacteria, spores etc. (that are larger than 0.5 μm) are available. Known as **membrane filters**, they can be fixed in funnel-like containers and autoclaved.

MEDIA PREPARATION

A culture medium contains essential minerals, and a carbon source (such as dextrose) usually supplemented with a complex proteinaceous material which provides vitamins, amino acids etc. The pH of the medium is adjusted to between 7.2 and 7.4 for most bacteria. The ingredients are taken in required quantities in water and transferred to culturing vessels. The medium, which is in a liquid form is known as a **nutrient broth** and the microbial culture grown in it as **broth culture**. Culture tubes or conical flasks can be used for raising broth cultures. The medium is made into a semi-solid gel by adding **agar agar** or gelatin to it prior to sterilization. Agar becomes a **sol (liquid state)** when boiled and solidifies into a **gel** on cooling to room temperature. The medium is tranferred to Petri dishes after sterilization. On cooling, such an agar medium provides a firm surface for the growth of microbial colonies.

Microbial cultures are maintained on agar slants for saving space and materials and for storing the cultures for extended periods. An agar slant is prepared by first boiling the medium containing agar. It is then transferred to culture tubes, which are then closed with cotton wool or a metal lid, and sterilized. After being taken out of the autoclave, they are kept in a slanting position until the agar sets into a gel. The slant provides a larger surface for the growth of microorganisms.

Enrichment Cultures

Microorganisms which are difficult to isolate, either due to very low population levels or special nutrient requirements are brought into culture by the **enrichment culture technique**. Enrichment culture technique is employed to build up populations of desirable organisms in natural habitats. A nutrient substance on which an organism grows well is added to the substrate such as soil, water, infected tissues etc., so that the population of the particular organism increases and the chances of its isolation improve. The use of dried blood for pathogens, and cellulose for cellulose decomposers are some such examples. Addition of chitin to soil was successfully employed to increase the populations of antagonistic actinomycetes for the control of some soil-borne diseases of plants.

ISOLATION AND INOCULATION

The minute size of microorganisms is compensated for by their very large numbers. Samples from natural habitats such as soil, water and sewage, harbour millions of microorganisms. Samples collected are diluted with water, and/or nutrient medium prior to isolation. Pathogenic microorganisms from infected tissues are collected by either inserting a swab (a thin rod with its tip covered with cotton), or aspirating into a syringe through its needle. The swab is shaken in the liquid to dilute the cells so as to facilitate isolation in pure culture. There are two general methods by which pure cultures are obtained:

Streak plate method. The diluted sample is streaked with an inoculation needle towards one side of a Petri dish to form a straight line on the surface of the medium. Another streak is made from the distal end of the first streak and at right angles to it. A third and a fourth streak are made in such a way that they do not touch each other midway. As the streaks are drawn, bacterial cells separate out and distinct, separate colonies develop. The cells from a discrete colony are picked up and transferred to an agar slant. Figure 67 illustrates the appearance of colonies on a streak plate.

Single cell isolation. The sample is examined under a counting chamber to determine the density of cells. It is then diluted to the extent that theoretically, one cell is distributed in two drops of liquid. Separate drops of liquid are placed on an agar medium. Discrete colonies that develop are subcultured on agar slants. Most of the microorganisms can be brought into pure culture by these procedures. Special methods have been developed to isolate microorganisms that are difficult of culturing, or for isolating genetically pure strains starting from single cells.

DIRECT OBSERVATION AND STAINING TECHNIQUES

Direct examination of unstained preparations is carried out by taking a drop of suspension in a cavity slide so that bacterial movement can be observed under a microscope. Being very small, bacteria show **Brownian**

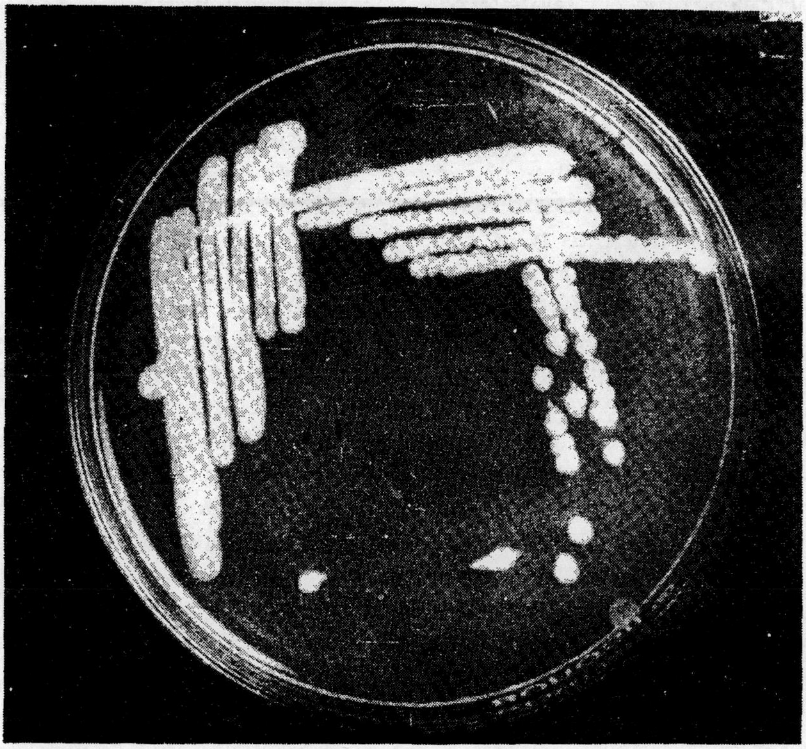

Fig. 67 Photograph of a streak plate.

movement, which is a small and rapid vibration in a liquid. Those having flagella also show **active movement** which can be easily distinguished from Brownian motion. **Hanging drop preparation** is also used for observation. A drop of suspension is placed on a coverslip. It is inverted on a glass ring placed on a slide. This device shows the movements more clearly. Unstained preparations are sometimes observed under a **phase contrast microscope**, which highlights different parts of cells. It is mainly employed for observing eukaryotic cells (algae, fungi and protozoa).

Staining procedures are adopted to improve the contrast, for observation either in diseased tissues or after bringing them into pure culture. The cells are first fixed so that internal and external features of cells remain in fixed position during staining. Fixation is done either by physical or by chemical means. A smear is made by taking a drop of suspension near one edge of a clean glass slide and drawing it towards the centre with the edge of another slide. It is then heated gently over a small flame until the water evaporates. Chemicals commonly used for chemical fixation are ethanol, acetic acid, glutaraldehyde etc.

Stains are dyes which have a **chromophore** (colouring component). They get attached to different parts of the cell so that surface features and internal parts become distinguishable. The dyes are either **basic dyes** such as methylene blue, crystal violet or malachite green, which colour negatively charged substance (cell walls, nucleic acids etc. of bacteria) or **acid dyes** such as eosin, acid fuchsin and rose Bengal, which bind to positively charged structures. When a single stain is used, the procedure is called **simple staining** and when a combination of stains is used it is **differential staining.** The latter highlights different components of the cell.

The most important differential staining is Gram's staining, which separates bacteria into two groups namely **Gram-positive** and **Gram-negative** bacteria. It was first developed by Christian Gram in 1884, and is widely used even now. Later it was shown that the two groups have different cell wall structure and hence show two types of reaction. The bacterial smear is first stained with crystal violet (primary stain). The smear is then treated with a mordant made up of iodine in potassium iodide solution. It helps in strong adherence of the stain. The smear is then de-stained in ethanol or acetone. Gram-positive bacteria retain the crystal violet dye while Gram-negative bacteria lose it. In order to stain this group, a basic dye, safranin is used for **counterstaining.** Gram-positive bacteria appear dark blue or purple while Gram-negative ones appear reddish or pinkish.

Another differential staining that is employed for *Mycobacterium* and a few other actinomycetes is **acid-fast staining**. These organisms are stained by adding basic fuchsin and phenol and heated. They are de-stained in a mixture of acid and alcohol. Those organisms that retain the stain are said to be acid-fast.

Special staining methods have been developed to distinguish **capsules** and **endospores**. Bacterial flagella are very fine structures, visible only in electron micrographs. However, their presence can be observed by adding a **mordant** which forms a coating around the flagella and takes up the stain. The mordants used are tannic acid or potassium alum and the stains used are basic fuchsin or rosaniline. The number and location of flagella on the cell can be seen after staining.

MAINTENANCE AND PRESERVATION OF CULTURES

Once a pure culture is obtained, it is maintained for further study. Actively growing agar slant cultures in small culture tubes or screw-capped vials are wrapped in polythene sheet or aluminium foil and kept in a refrigerator so that evaporation is reduced and metabolic activity remains at a low ebb due to reduced aeration. The microorganism from the stored culture can be recovered by inoculating freshly prepared agar slants, usually within 2-3 months. For long-time preservation, the actively growing cultures in tubes or vials are **overlaid with sterilized paraffin oil** so as to cover the entire agar surface. As paraffin oil is neutral and non-evaporating, the culture retains viability for several months.

Lyophilization of cultures preserves them for much longer periods of upto several years. Lyophilization consists of taking nutrient broth culture in small vials and attaching their mouths to suction tubes in a lyophilizer.

The moisture is quickly evacuated or **sublimed** under high vacuum, so that temperature does not rise. The mouths of the vials are then sealed prior to removing them from the lyophilizer so that vacuum is maintained in the vial.

Besides the basic methods, numerous biochemical and biophysical techniques in the study of microbial taxonomy, metabolism, genetics and ecology including the use of computers and automation, have greatly advanced the knowledge and application of microorganisms for human welfare.

LIST OF BOOKS FOR FURTHER READING

Alexopoulous, C.J. and Mins, C.W., *Introductory Mycology*, 3rd ed., John Wiley, New York, 1979.

Atlas, R.M. and Bartha, R., *Microbial Ecology*, Benjamin-Cummins Publishing Co., Redwood City, California, 1993.

Benn, R.A.V., *Aids to Microbiology and Infectious Diseases*, Churchill-Livingstone, London, 1985.

Benson, H.J., *Microbiological Applications*, 6th ed., Wm. Brown & Co., New York, 1994.

Bold, H., *Introduction to Algae*, Prentice Hall, Englewood Cliffs, New Jersey, 1985.

Brock, T.D. and Madigan, M.T., *Biology of Microorganisms*, 6th ed., Prentice Hall, Englewood Cliffs, New Jersey, 1991.

Creuger, W. and Creuger, A., *Biotechnology—A textbook of industrial microbiology*, Science Tech. Inc., Madison, 1984.

Fencher, T., *Ecology of Protozoa*, Springer-Verlag, Berlin, 1987.

Frazier, W.C. and Westhoff, D.C., *Food Microbiology*, Tata McGraw-Hill, New Delhi, 1978.

Freifelder, D., *Essentials of Molecular Biology*, Narosa Publishing House, New Delhi, 1985.

Gottschalk, G., *Bacterial Metabolism*, 2nd ed., Springer-Verlag, Berlin, 1986.

Hyde, R.M. and Pathode, R.A., *Immunology*, John Wiley, New York, 1987.

Jay, J.N., *Modern Food Microbiology*, 3rd ed., Van Nostrand, New York, 1978.

Kotpal, R.L., *Protozoa*, Rastogi Publications, New Delhi, 1980.

Mandahar, C.L., *Introduction to Plant Viruses*, 2nd ed., S. Chand & Co., New Delhi, 1984.

Marx, J.L., *A Revolution in Biotechnology*, Cambridge University Press, Cambridge, 1989.

Matthews, R.E.F., *Plant Virology*, 3rd ed., Academic Press, New York, 1991.

Pelczar, M.J., Chan, E.C.S. and Kreig, N.R., *Microbiology*, McGraw-Hill, New York, 1993.

Prescott, L.M., Harley, J.P. and Klein, D.A., *Microbiology*, 2nd ed., Wm. C. Brown, New York, 1990.

Rangaswamy, G. and Bagyaraj, D.J., *Agricultural Microbiology,* Prentice-Hall of India, New Delhi, 1993.

Reed, G. (Ed.), *Prescott and Dunn's Industrial Microbiology,* CBS Publications, New Delhi, 1987.

Roitt, I., Brostoff, J. and Male, D., *Immunology,* 3rd ed., Mosby, New York, 1993.

Satoskar, R.S. and Bhandarkar, S.D., *Pharmacology and Pharmacotherapeutics,* 13th ed., Popular Prakasan, New Delhi, 1993.

Voit, D. and Voit, J.G., *Biochemistry,* John Wiley, New York, 1990.

Webster, J., *Introduction to Fungi,* 2nd ed., Cambridge University Press, Cambridge, 1980.

INDEX

A

Acetate, 77, 153
Acetate fermentation, 88
Acetoin, 86
Acetoacetyl CoA, 88
Acetobacter, 88, 147, 153
A. suboxidans, 153
Acetogenic bacteria, 89
Acetone, 152
Acetyl CoA, 77, 88
Achromobacter, 82, 161
Acid-fast staining, 188
Acid fuchsin, 182
Acquired immunodeficiency syndrome (AIDS), 55, 174, 178
Acquired immunity, 132
Acridine dyes, 110
Actinomadura, 40
Actinoplanes, 33, 39
Actinomycetes, 20, 30, 33, 38, 39
Activated sludge, 143
Activation energy, 68
Active
  immunity, 132
  site, 68
  transport, 64
Adenophosphosulfate (APS), 91
Aerobacter aerogenes, 147, 154
Aerobes, 102
Aerobic,
  chemolithotrophic, 37
  Gram-negative cocci, 32
  Gram-negative rods, 32
Aflatoxin B1, 110
Aflatoxins, 164
Agar-agar, 185
Agaricus, 26, 27, 60
A. bisporus, 60
Agglutination, 134
Agrobacterium, 32, 34
A. tumefaciens, 34, 123
AIDS, 174, 178
Aldrin, 143
Akinetes, 37
Algae, 23, 24, 100, 136, 142, 187
Alcoholic beverages, 145
Alcaligenes, 160, 161, 163
A. eutrophus, 76, 102
A. faecalis, 82
Aliphatic hydrocarbons, 99
Allergies, 133
Allosteric property, 69
Alternaria, 26, 28, 162
Amino acids, 89, 90, 98, 154
Aminoglycosides, 157
Amoebic dysentery, 166, 175, 176
6-Amino penicillanic acid, 154 156
Ammonia, assimilation of, 90, 106
Ammonification, 138
Amoebae, 22
Amoeba proteus, 22
Amphotericin B, 175
Ampicillin, 180
Anabaena, 33, 37, 105
Anabolic process, 73
Anaerobic
  bacteria, 65, 166
  digestion, 88, 144
  respiration, 82
Anaplerotic sequences, 81, 82
Antenna molecules, 73
Anthrax, 1, 165, 178
Animal viruses, 44, 45, 52
Antibiotics,
  chemical structures of, 156
  mode of action of, 180
  production of, 38, 145, 151, 155, 179
Antibodies,
  monoclonal, 48, 122, 125, 130, 133
  polyclonal, 133
Anticodon, 112
Antigenic properties, 48, 49, 124, 126, 133
Antigen
  receptors, 129
  recognition, 126

Antimicrobial substances, 175, 180
Antiseptics, 179
Antiserum, 49, 133
Antitoxin, 173
Apoenzyme, 68
Aquaspirillum, 32, 35, 103
Aquatic habitats, 142
Archaebacteria, 17, 20, 33, 40
Archaebacteria, cell wall less, 33, 41
Archaeolobus, 41
Aromatic compounds, utilization of, 99
Arthrobacter, 32, 36, 76
A. simplex, 158
Ascomycetes, 26
Aspergillus, 26, 60, 140, 151, 153, 162, 166
A. flaves, 164
A. niger, 147, 153, 154
A. ochraceus, 164
A. parasiticus, 164
A. oryzae, 84, 154
A. wentii, 147, 154
Asthma, 133
Atheletes foot, 174
ATP formation, 79
Attenuation, 132
Autoimmune diseases, 133, 170
Autotrophs, carbon assimilation in, 103
Auxotrophs, 108, 109, 110
Azoferredoxin, 106
Azospirillum, 32, 35, 105
Azotobacter, 2, 32, 34
A. agilis, 105
A. chroococcum, 76, 105

B

B-cells, 126, 127, 129, 130
Bacilli (Singular-bacillus), 29
Bacillus. 11, 15, 29, 36, 97, 140, 143, 151, 154, 160, 161, 162, 163
B. amyloliquifaciens, 154
B. anthracis, 1, 36, 165
B. cereus, 36, 76, 154, 163, 168, 169
B. coagulans, 154, 163
B. licheniformis, 82, 147, 154, 157
B. megaterium, 140
B. polymyxa, 147, 157
B. subtilis, 76, 98, 147
B. thuringenesis, 36
Bacitracin, 36
Back mutation, 111
Bacteria, 20, 32, 34, 73, 179
  conjugation in, 132
  magnetotactic, 103
Bacteriochlorophylls, 72
Bacteroids, 106
Bacteroides, 97, 166, 167
Bakers yeast, 151
Bacteriophages, 3, 42, 45, 56, 119
Bacteriostatic, 157
Basal body, 13
Base
  deletion, 110
  insertion, 110
  substitution, 110
Basophils, 124
Batch cultures, 67
Batch fermentation, 148, 150
Bdellovibrio, 32, 35
B. bacteriovorus, 35
Beijerinckia indica, 105
Beer making, 84, 148
Benign tumors, 60
Beggiatoa, 38
Benthic zone, 142
Bifidobacterium, 86
B. bifidum, 86
Bifidum pathway, 86
Binary fission, 15
Binal symmetry, 43, 58, 59
Biodegradable, 143
Biofertilizers, 140
Biogeochemical cycles, 137
Biological
  nitrogen fixation, 103
  oxygen demand (B.O.D.), 143
Biomass, 66
Biomining, 141
Biosynthesis, 89
Biotin, 64, 154
Blastomycosis, 175
Blastomysis dermatitis, 175
Blood groups, 133, 134
Bone and Joint tuberculosis, 171
Bone marrow stem cells, 126
Bordetella pertussis, 165, 171
Botulism, 163
Botulitoxin, 163
Bradyrhizobium, 32, 34, 105, 107
Branhamella, 167
5–Bromouracil, 110
Brownian movement, 186-187
Broth culture, 185

Brucella abortus, 81
Budding, 16, 49
Burkitts lymphoma, 60
Burst size, 49
n-Butanol, 150, 152
Butyrate, 87, 88
Butyrivibrio, 87

C

Cadang-cadang, 17
Calvin cycle, 101, 103, 104
Campylobacter, 83
C. jejuni, 167
Candida, 153, 160, 162, 166, 175
C. albicans, 175, 180
C. utilis, 147
C. lipolytica, 99
C. tropicalis, 158
Candidiasis, 175
Canning, 161, 163
CAP protein, 120
Capsid, 42
Capsule, 13
Carbon cycle, 138
Carbohydrates, 93
Carbon sources, 61
Carboxysomes, 11, 103
Carcinogens, 110
Carotenoids, 73
Carrier
  proteins, 64, 93
  substances, 70
Catabolism, 73
Cell division, 15
Cellobiose, 97
Cells of immune system, 124
Cellular compounds, biosynthesis of, 89
Cellular immunity, 124
Cellulomonas, 97
Cellulose, 98, 136, 145, 148, 167
Cell wall, 8
Cephalosporin, 155, 180
Cephalosporium, 151
C. acremonium, 147, 157
C. roseum, 99
Chamberland filters, 2
Cheese making, 154
Chemical preservatives, 162
Chemiosmotic hypothesis, 80
Chemostat, 67, 149, 150
Chemotherapy, 171, 179
Chicken pox, 178
Chitin, 97, 98
Chlamydia trachomatis, 18, 32, 35
Chlamydiales, 18, 35
Chlamydomonas, 24
Chlorobiaceae, 100
Chlorobium, 33
Chloroflexus, 33
Chloroplasts, 13, 103
Cholera, 172, 173
Cholera toxoid, 173
Chromatiaceae, 100, 103
Chromatin, 12
Chromatium, 15, 33
Cilia, 15
Ciliates, 22
Ciliophora, 22
Cis-aconitate, 77
Cistron, 111
Citrate, 82, 86, 145, 153
Citrobacter, 32, 34
Cladophora, 24, 25
Claviceps, 151
C. purpurea, 164
Clostridium, 11, 32, 36, 87, 89, 97, 98, 152, 158, 160, 161, 163, 167, 169
C. acetobutylicum, 88, 147
C. botulinum, 163, 165, 173
C. butylicum, 163
C. formicoaceticum, 89
C. kluyveri, 87
C. nigrificans, 163
C. pasteurianum, 87, 105, 106, 163
C. perfringens, 173
C. putrifaciens, 163
C. sporogenes, 163
C. tetani, 36, 129, 165, 173
C. thermoaceticum, 89
C. thermosaccharolyticum, 163
Cloning vector, 56, 59, 121
Cloxacillin, 180
Coat protein, 56
Coccidiomycosis, 175
Codaeum, 24
Coenocytic filaments, 23
Coenzyme, 68
Colicins, 115
Coliforms, 166
Coliform test, 173
Col-plasmids, 115
Colony characters, 30
Complement system, 125

Compound microscope, 181
Conjugation in bacteria, 118, 151
Continuous culture, 148
Conditional mutants, 110
Corn, 145, 153
Corynebacterium glutamicum, 147, 154
C. diphtheriae, 36, 129, 130, 165, 170
Cortisone, 158
Cosmid, 59
Coxiella, 32, 35
Coccidioides immilis, 175
Cristae, 13
CRP proteins, 125, 129
Crystal violet, 182, 188
Cumulative mechanism, 69
Curing, 114, 163
Curvularia lunata, 158
Cyanobacteria, 20, 21, 30, 33, 37, 100, 136, 142
Cyclic AMP, 120
Cytokines, 125, 127, 129, 131
Cytoplasmic matrix, 10, 12
Cytophaga, 33, 38, 97
Cytoskeleton, 12
Cytotoxins, 168

D

Dental caries, 166
DDT, 143, 176
Deep mycoses, 175
Defective phages, 118
Degenerate code, 110, 111
Degradation of polymers, 97
Delayed hypersensitivity, 133
Denitrification, 82
Denitrifying bacteria, 82
Dermatophytes, 175
Derxia gummosa, 105
Desulfobacter, 83
Desulfococcus, 83
Desulfolobus, 83
Desulfotomaculatum, 141
Desulfomonas, 35, 83
Desulfosarcina, 83
Desulfovibrio, 32, 35, 83, 141
Desulfuromonas, 35, 83
Detergents, 143
Diacetyl, 86
Diaminopimelic acid, 154
Diarrhoea, 172
2, 4-Dichlorophenoxy acetic acid, 143
Dictyosomes, 13
Dictyota dichotoma, 25
Diphtheria, 170, 179
  toxin, 129, 170
Dipicolinic acid, 11
Diseases
  autoimmune, 133, 170
  of Animals, 52
  of Humans, 52–55, 163–174
  of Plants, 51
Diseases caused by
  bacteria, 168
  fungi, 174
  protozoa, 175
  virus, 132
Disinfectants, 179
DNA-Cos hybrid, 59
DNA ligases, 121
  polymerization, 95
  replication of, 95
  splicing, 121
  synthesis, 94
Donor cell, 118
Double helix, 94, 95
DPT vaccine, 170, 171, 179
Drug resistence plasmids, 113

E

Eclipse period, 49
Eczema, 133
ED pathway, 74, 75, 76, 84
Electron
  acceptor, 69, 70
  carrier, 71
  donor, 69, 70
  transport, 74, 83
ELISA technique, 134, 135
EMP pathway, 74, 75, 76, 84
Endemic, 177
Endoplasmic reticulum, 12
Endospore forming bacteria, 11, 12, 32, 36
Endospores, 11, 36, 188
Endothia parasitica, 159
Endotoxins, 168, 169
Entamoeba histolytica, 21, 166, 176
Enrichment culture, 185, 186
Enteric bacteria, 171
Enterobacteriaceae, 34
Enterococcus, 32, 36

Enterotoxin, 130, 168 169, 173
Enzymes and Enzyme Groups, 68, 69, 71, 154
- Acetaldehyde dehydrogenase, 84
- Acetate kinase, 86
- Acetyl-CoA carboxylase, 93
- ACP dehydratase, 93
- ACP reductase, 93
- ACP synthetase, 93
- ADP-glucose pyrophosphorylase, 94
- Alcohol dehydrogenase, 84
- Aminoacyl-tRNA synthetase, 96
- Aminotransferase, 94
- Amylases, 97, 145, 152, 154
- APS phosphokinase, 91
- Aspartatetranscarbamylase, 91
- ATPase (ATP synthetase), 80
- ATPsulphurylase, 91
- Cl enzymes, 97
- Cχ enzymes, 97
- Catalase, 83
- Cellobiose, 97
- Cellulobiose phosphorylase, 97
- Cellulases, 97
- Chitinases, 97
- Cis-aconitate hydratase, 77
- Citrate synthetase, 77
- Dihydrolipoic dehydrogenase, 77
- Dihydrolipoic transacetylase, 77
- DNA ligases, 121
- DNA polymerases, 95
- DNA-dependent RNA polymerase, 95
- DNAses (Deoxyribonucleases), 98
- Endogenous transcriptase, 54
- Enolase (Phosphopyruvate hydratase), 74
- Epimerase, 93
- Fumarase, 71, 78
- Fructose-1, 6-biphosphate phosphotase, 104
- Fructose biphosphate aldolase, 74
- β-Galactosidase, 120
- Galactokinase, 77
- Glocose: galactose-l-phosphate uridyltransferase, 77
- Glucose isomerase, 154
- Glucose-6-phosphate dehydrogenase, 76, 80
- Glutamate dehydrogenase, 91
- Glutamate synthetase (GOGAT), 91
- Glutamine synthetase (GS), 91
- Glyceraldehyde-3-phosphate dehydrogenase, 74, 81, 104
- Glycogen synthetase, 94
- Hydrogenase, 88, 106
- Isoamylases, 154
- Isocitrate dehydrogenase, 78
- 3-Ketoacyl ACP reductase, 93
- Lactonase, 80
- Lactose permease, 77
- Lysozyme, 49
- Malonyl-transacelase, 93
- Maltose phosphorylase, 97
- Neuraminadase, 54
- Nitrate reductase, 82
- Nitric oxide reductase, 83
- Nitrogenase reductase (Azoferredoxin), 106
- Nitrogenase (Molybdoferredoxin), 106
- PAPS reductase, 91
- PEP carboxylase, 69, 81
- Phosphoglucomutase, 93
- Pectinases, 97
- Penicillin acylase, 154
- 6-Phosphogluconate aldolase (KDPG aldolase), 75, 76
- 6-Phosphogluconate dehydratase, 76
- 6-Phosphogluconate dehydrogenase, 80
- 3-Phosphoglycerate kinase, 74
- Phosphoglycerate mutase, 74
- Phosphoketolase, 86, 87
- Phosphoribose isomerase, 80
- Phosphoribulokinase, 103
- Phosphotransacetalase, 86, 88
- Pyruvate decarboxylase, 84
- Pyruvate dehydrogenase, 69, 77
- Pyruvatekinase, 74
- Restriction endonucleases, 121
- Rennet, 159
- Replicase, 56
- Reverse trascriptase, 55
- Ribose-5-phosphate-3-epimerase, 80
- Ribozyme, 55
- Ribulose-1, 5-biphosphate carboxylase, 103, 104
- RNA polymerase, 95, 112
- RNAses (Ribonucleases), 98
- Succinate dehydrogenase, 79
- Succinate thiokinase, 78
- Superoxide dismutase, 65, 83
- Transaldolase (TA), 80

Transketolase (TK), 80
Triosephosphate isomerase, 74, 104
UDP glucose epimerase, 77
UDP-glucose pyrophosphorylase, 93
Eosinophils, 124
Epidemics, control of, 178
Epidemiology, 177
Epidermophyton, 174
Episomes, 113
Epitopes, 125, 126, 129
Ergotism, 164
Erwinia, 34, 97, 160, 162
E. amylovora, 84
Erysiphe, 26
Erythromycin, 157
Escherichia, 32, 160
E. coli, 14, 34, 56, 58, 76, 79, 113, 115, 116, 118, 119, 120, 130, 147, 151, 154, 158, 163, 165, 168, 173
E. coli (genetically engineered), 147
Ethanol, 85, 152
fermentation of, 84
Eubacteria, 17, 20
Eubacterium limosum, 87
Euglena, 24
Eukaryotes, 15
European encephalitis, 52
Exotoxin, 168, 169, 172
Extreme
halophiles, 41, 66
thermophiles, 41

F

F-factor, 115, 118
F plasmid, 116
Fab portion, 129
Facilitated diffusion, 63
Facultative anaerobes, 65, 166
Faenia, 38
Fatty acid synthesis, 92
FC region, 129
Feedback mechanism, 69
Fermentation products, 146, 152
Fermentations,
Acetate, 88
Bacterial, 84
Butanol-acetate, 87
Butyrate, 87
Ethanol, 84
Methane, 88
Ferredoxin, 72, 106
Fertility factor, 115
Fimbriae, 13
Fish, exploited for, 142
Flagella, 13
Flagellates, 22
Flagellin, 14
Flavobacterium, 143, 160, 161
Flavodoxin, 72
Flavoproteins, 71
Fluorescent antibodies, 134
Food
intoxication, 163
poisoning, 172, 173
preservation, 162
spoilage, 159, 160
Foot and mouth disease, 166
Foraminifera, 22
Frankia, 33, 105, 107
Fritschiella, 24
Fucus, 24
Fumarate, 78, 83
Fungal phages, 60
Fungi, 23, 26, 123, 160, 174, 175 179, 187
Furious rabies, 53
Fusarium, 26, 160, 162, 164
F. moniliforme, 147, 158
Fusobacterium nucleatum, 87

G

Gal operon, 115
Galactose, 77
Gallionella, 103
Gas gangrene, 173
Gastro-intestinal tract, 163, 166, 171
Gas vacuoles, 11
Gene
amplification, 117
cloning, 121
library, 122
Genes, 111
overlapping of, 48
Generation time, 66
Genetic code, 111, 112
Genetic engineering, 108, 152
Gentamicin, 157
Geodermatophilum, 33, 39
Gibberella fujikuroi, 158
Gibberellins, 158
Gliding bacteria, 38

Gloeocapsa, 33, 37
Glomus, 26
Gluconobacter, 88
Glycoprotein, 53
Glucose, 74, 77
Glucose-6-phosphate, 74, 75, 76
Glutamine, 91, 107
Glyceraldehyde-3-phosphate, 75, 76, 87
Glycocalyx, 94
Glycogen, 11, 98
Glycolysis, 74
Glyoxylate cycle, 82
Gobar gas, 144
Golgi apparatus, 13
Gonorrhoea, 174, 178
Gram-negative bacteria, 8, 32
Gram-positive bacteria, 8
Gram's staining, 8, 188
Grana, 13
Green bacteria, 100
Granulocytes, 124
Griseofulvin, 175
Group transport, 64
Growth factors, 63
Growth, measurement of, 66
Gymnodinium, 24

H

H-antigens, 171
Haemagglutination, 134
Haemocytometer, 182
Haemolysis, 168
Hairycell leukaemia, 60
Half-life, 143
Halobacterium, 33, 41
H. salinarium, 41
Hansenula, 160
Halophiles, 66
Haptens, 124
Heat labile, 168
Heat stable, 168
Helical symmetry, 42
Helminths, 131
Hepatitis B, 60
Heterotrophs, 61, 136
Hepatocellular carcinoma, 60
Herd immunity, 178
Heterocysts, 37
Heterofermentative pathway, 86, 87
HFr strain, 115, 118
High fructose syrups, 145
Histoplasma, 175
H. capsulatum, 175
Human disease, 165
Human growth hormone, 122
Human T-cell viruses, 60
Humoral immunity, 124
Humus, 137
Hybrid DNA, 59
Hybridoma, 133
Hydrogen gas, 101, 163
Hydrogen-oxidizing bacteria, 102
Hydrophobia, 53
Hydrothermal vents, 142
Hydroxymethyl cytosine, 58
Hypersensitivity, 133
Hyphomicrobium vulgare, 82
Hypolymnion, 142

I

Icosahedral symmetry, 43
Immobilization of enzymes, 155
Immunity
  acquired, 132
  active, 132
  natural, 124
  passive, 132
Immunity to
  microorganisms, 129
  protozoan, 131
  viruses, 129, 131
Immunoelectrophoresis, 134
Immunogens, 124
Immunoglobulins, 127
  IgA, 127, 128
  IgD, 127, 129
  IgE, 127, 129
  IgG, 127, 128, 132
  IgM, 127, 128
Immunological memory, 126
Immunology, 2, 124
Inclusion bodies, 11
Insertion elements, 116
Insulin, 59, 122
Interferon, 59, 123, 125, 131
Interleukens, 126
Iron-oxidizing bacteria, 102
Isocitrate, 77, 78
Isoenzymes, 68
Isolation of microorganisms, 186

K

K-antigens, 171
Kanamycin, 157
Klebsiella, 32, 34, 171
K. pneumoniae, 105, 106, 167
Kluyveromyces fragilis, 147, 152
Koch's postulates, 1
Kuru disease in humans, 17

L

Lactic acid bacteria, 36, 159
Lactobacillus, 36, 86, 159, 160
L. acidophilus, 159
L. delbruckii, 147, 153
L. bulgaricus, 36, 147, 153, 159
L. pentosus, 153
Lactococcus, 32, 36
Lactose, 77, 167
Lactose operon model, 77
β-lactum antibiotics, 155
Leavening of dough, 84
Leghaemoglobin, 72, 73
Leprosy, 129
Leptospirillum, 141
Leuconostoc, 32, 36, 86, 159, 160, 162
L. citrovorum, 159
L. dextronium, 159
L. mesinteroides, 159
L. glutamate, 91, 107
Leucothrix, 33, 38
LGHs, 125
Lipid envelope, 42
Lignin, 136
Lipid membrane, 52
Lipids, biosynthesis of, 92
Lipopolysaccharides, 8, 168
Lipoprotein envelope, 8, 42
Lithotrophs, 61, 102, 140, 141
Looped rolling circle mechanism, 115
Lymphocytes, 125
Lymphogranuloma venereum, 18
Lymphokines, 125
Lyngbya, 33, 37
Lyophilization, 188
Lysergic acid, 164
Lysine, 154
Lysis protein, 56
Lysogenic phages, 58
Lysozyme, 11, 49

M

Macrolide, 157
Macronutrients, 61
Macrophages, 129, 130
Madurella mycetoma, 174
Maduromycetes, 40
Magic bullet, 179
Maintenance of cultures, 188
Malachite green, 188
Malaria, 166, 175
Malate, 77, 78, 82
Mammalian cell lines, 123
Mastigophora, 22
Madura foot, 174
Maturation protein, 56
Measles, 178
Meat-bacteria colonizing, 160
Media-preparation, 185
Meiosis, 16
Membrane filters, 185
Memory cells, 126
Menaquinone, 71, 79
Meningitis, 171
Messenger RNA (mRNA), 95, 111
Meso-diaminopimelic acid, 8
Mesophiles, 65
Mesosome, 9
Metabolic plasmids, 114
Metachromatin granules, 11
Methane, 88, 138, 144
Methanobacterium, 33, 41, 88, 143
Methanococcus, 33, 88, 143, 144
Methanogenic archaebacteria, 40
Methanogens, 88
Methanopyrus kandleri, 41
Methanosarcina, 33, 41, 88, 143, 144
Methanospirillum, 33, 41, 88, 143, 144
Methylene blue, 182, 188
Methylococcus, 32, 34
Methylomonas methalonica, 32, 34, 158
Methylotrophs, 34
MHC molecules, 129
MHC proteins, 127, 131
Michaelis constant, 68
Microaerophils, 65
Micrococcus, 32, 35, 143, 160, 161, 163, 166
M. denitrificans, 82
Micromonospora, 33, 39
M. purpurea, 147, 157
Micronutrients, 61, 141

Microscopes,
compound, 181
light, 181, 182
scanning electron (SEM), 184
Microscope, transmission electron (TEM), 183, 184
Microscopic methods, 181
dark field, 182
electron, 183
fluorescence, 181
phase contrast, 182, 187
Microsporum, 174
Minimal medium, 109
Mini Mu phages, 122
Missense mutation, 110
Mitochondria, 6, 13
Mitosis, 16
Molasses, 152, 153
Molybdoferredoxin, 106
Mononucler phagocytes, 124
Monosodium glutamate, 145
Mosquitos, 178
Mucor, 26, 153, 160
M. pusillus, 154, 159
Multicomponent viruses, 48, 51
Murein, 8
Mutagens, 108, 109, 151
Mutants, 108, 109, 151
Mutator genes, 110
Mutations, 108, 110
Mycobacteria, 32, 36
Mycobacterium, 36, 143
M. tuberculosis, 37, 130, 157, 165, 167, 170, 171
M. leprae, 37
Mycoplasma, 18, 32, 35
M. mycoides, 35
Mycorrhiza, 25
Mycoses, 174
Mycotoxin problem, 163
Myeloid cells, 124
Myxobacteria, 38
Myxococcus, 38

N

N-acetylglucosamine, 8, 40
N-acetylmuramic acid, 8
Nasopharyngeal carcinoma, 60
Natural immunity, 124
Navicula, 24
Negative regulation, 120
Negotiable energy, 79
Neisseria gonorrhoae, 165, 174
Neomycin, 157
Neoplasia, 60
Neubar's chamber, 182
Neuraminadase, 53
Neurotoxins, 168
Neutralization, 133
Neutrophils, 124
Nif genes, 106, 107
Nitrate reduction, assimilative, 63, 83, 91
Nitrate reduction, dissimilative, 83, 91
Nitrification, 138
Nitrifying bacteria, 37, 101
Nitrobacter, 33, 37, 38, 101, 139
Nitrogen cycle, 138, 139
Nitrococcus, 38, 101
Nitrogen fixation, 63, 103, 105, 107, 138
Nitrogen-fixing organisms, 105
Nitrogen fixation and assimilation, 107
Nitrogenases, 63, 104
Nitrogenous fertilizers, 138
Nitrogen respiration, 82, 139
Nitrosococcus, 33, 38, 101, 139
Nitrosomonas, 33, 37, 101, 139
Nitrosospira, 33, 38
Nitrosovibrio, 101
Nitrospira, 101
Nocardia, 33, 38, 143
N. asteroides, 38
N. mediterranea, 147, 157
N. petroliphila, 99
NK cells, 125, 131
Nod genes, 107
Nodulins, 106
Nomenclature, bionomial system of, 5
Nonsense mutation, 110
Normal microflora of body, 166
Nostoc, 33, 37
Ntr gens, 106
Nucleic acids, 98, 140
Nuclear envelope, 12
Nucleocapsid, 42
Nucleoid, 10
Nucleolus, 12
Nucleoplasm, 12
Nucleotide, 92
Nucleotides, biosynthesis of, 91
Numerical taxonomy, 31

O

O antigens, 171
Obligate anaerobes, 167
Obligate lithotrophs, 140, 141
Oceanospirillum, 35
Ochratoxin, 164
Oncogenes, 60
On-off regulation, 121
Operator gene, 121
Opsonization, 125, 129, 130
Osmophiles, 65
Osmotic pressure, 65
Outer membrane, 94
Oxaloacetate, 78, 82
Oxalosuccinate, 78
Oxidation-reduction, 69
Oxidative phosphorylation, 79
Oxidizing agents, 180
α-oxoglutarate, 78, 82

P

Padina, 24
Paecilomyces, 153
Pandemic, 177
Paraaminobenzoic acid, 180
Paramoecium, 23
Parasites, 23
Paratyphoid, 172
Passive diffusion, 63
Passive immunity, 132
Pasteur effect, 84
Pasteurisation, 2, 159, 161, 178
Patulin, 164
Pectic substances, 97
Pectinases, 154
Pediococcus, 159
Penicillin, 3, 155, 156, 180
Penicillium, 26, 60, 151, 153, 156, 160, 162
P. camembertii, 159
P. chrysogenum, 147, 156
P. graminis, 26
P. griseofulvum, 147
P. patulum, 164
P. roqueforti, 159
Pentose phosphate cycle, 81
Peptide antibiotics, 157
Peptidoglycan, 8, 94
Periplasmic space, 8
Perishable foods, 159
Pernicious anemia, 133
Pertussis, 171, 179
Peziza, 26
Phagocyte, 124, 125, 131
Phagocytosis, 124, 130, 131
Pharyngitis, 130
Phialophora, 174
Phosphoroclastic reaction, 87
Phosphobacterin, 140
Phosphoenolpyruvate (PEP), 74
3-Phosphoglycerate, 74
6-Phosphogluconate, 76
Phospholipids, 92, 93, 140
Phosphorylation,
  electron transport, 79
  oxidative, 79
  substrate level, 74
Phosphorus cycle, 139, 140
Photoautotrophs, 61
Photobacterum, 32, 35
Photoorganotrophs, 61
Photosynthetic
  bacteria, 33, 37
  pigments, 72
Photosystem I, 100
Photsystem II, 100
Phycobiliproteins, 73
Phytin, 140
Phytophthora, 26
Phytoplankton, 142
Pickling, 161
Pili, 44
Pinocytosis, 21, 44
Pityrosporum ovalis, 166
Plague,
  bubonic, 172
  pneumonic, 172
Plant viruses, 44, 51
Plaques, 50
Plasma
  cells, 126
  membrane, 94
Plasmids, 11, 108, 113, 116
Plasmids,
  conjugative, 113
  drug resistance (R), 113
  virulence, 114
Plasmodiophora brassicae, 28
Plasmodium, 21, 28, 131, 166, 175
P. vivax, 166
Plasmopara viticola, 26

Plastids, 6
Pleomorphic, 18, 30, 43
Plug-flow method, 149, 150
Pleuropneumonia, 18
PMN cells, 124
Pneumonia, 168, 178
Point mutation, 110
Polio, 166
Pollution, 142
Polycistronic, 51
Polyclonal antibodies, 133
Polychlorinated biphenyls, 143
Poly-β-hydroxybutyrate, 11, 94
Polymorphs, 125
Polymyxin, 36
Polynucleotides, 94
Polypeptide, 96, 112
Polyphosphate, 94
Polyprotein, 52
Polyribosomes (Polysomes), 112
Polysiphonia, 24
Porphyrin ring structure, 72
Positive regualtion, 120
Potato blight, 165
Precipitation reaction, 134
Preservation of cultures, 188
Primary electron donor, 70
Primary producers, 136
Prions, 17
Probes, 122
Prochloron, 21
Prochlorophyta, 21, 72
Prochlorothrix, 21
Prokaryotes, 5, 7
Prokaryotic cells, 7
Prophage, 58
Propionibacterium, 155, 166
P. shermanii, 147
P. acnes, 166
Prosthetic groups, 68
Promoter, 111, 120
Proteases, 154
Protein synthesis, 95, 96
Proteus, 34, 171
Protists, 5
Proton motive force, 80
Prototroph, 108, 109
Protozoa, 21, 136, 175, 179, 187
Pseudomonas, 14, 32, 34, 58, 97, 115, 130,
  140, 143, 151, 160, 161, 163
P. aeruginosa, 34, 177
P. amylodesmosa, 154
P. carboxyvorans, 102
P. denitrificans, 82, 155
P. flava, 102
P. fluorescence, 34, 76, 99
P. methylotrophus, 158
P. phaseolicola, 60
P. putida, 144
P. saccharophila, 75, 76
P. stutzeri, 82
Pseudoparenchymatous, 23
Psuedopodia, 21
Psychrophiles, 65, 142
Puccinia graminis, 26
Pure culturing, 181
Purple bacteria, 100
Putrefaction, 160
Pyogenic, 177
Pyricularia oryzae, 26
Pyruvate, 77

Q

Q-fever, 18
Quinone, 71
Quick freezing, 161

R

Rabies, 53
Rabies vaccine, 2, 53
Reaction centre molecules, 73
Recalcitrant chemicals, 143
Recepient, 118
Receptive site, 44
Recombinant DNA, 108, 121
Recombinations in
  bacteria, 118
  bacteriophages, 119
Redox potentials, 70, 160
Regulation of gene action, 120
Renin, 154
Rennet, 159
Replenishing sequences, 81
Replica plating, 109
Replication, 42, 49, 95, 116
Replicative form, 48
Respiration, 73, 82
Respiratory tract, 167
Reverse transcriptase, 55
Reversion, 111
RH factor, 134

Rheumatic fever, 170
Rheumatoid arthritis, 133
Rhizobium, 2, 32, 34, 105, 106, 107
R. japonicum, 76
Rhizopus arrhizus, 147
R. nigricans, 147
R. niveus, 147
Rhodophyta, 23
Rhodopseudomonas, 33
Rhodospirillaceae, 100
Rhodospirillum, 33, 105
R. rubrum, 101
Rhodotorula, 160, 162
Ribosomal RNA (rRNA), 95
Ribosomes, 111
Ribose-5-phosphate, 80
Ribulose-5-phosphate, 80
Rickettsias, 18, 32, 35
Rifamycin, 171
Ringworm, 174
RNA-synthesis of, 94
mRNA, 112, 113
Rocky mountain spotted fever, 18
Ruminococcus, 97

S

Sabin vaccine, 52
Saccharomyces, 26, 84, 162
S. cerevisiae, 84, 145, 147, 152
S. lipolytica, 158
Saccharomycopsis, 158
Safranin, 188
Salmonella, 32, 34, 60, 115, 130, 160, 163, 168, 169, 171
S. enteritidis, 165
S. hirschfeldii, 172
S. newportii, 172
S. paratyphi, 165, 168, 169, 172
S. schotimulleri, 172
S. typhi, 165, 168, 169
Saprolegnia, 26
Sarcina ventriculi, 84
Sarcodina, 22
Sauerkraut, 87
Scrapie disease, 17
Sebaceous glands, 166
Segmented genome, 48, 53
Semisynthetic penicillins, 156
Serology, 133, 135
Serology of viruses, 48
Septicaemic, 172
Serratia, 160, 161
Sewage, microorganisms in, 143
Sex pilus, 118
Sexually transmitted diseases (STD), 173, 178
Shigella, 32, 34, 115, 168, 169, 172
S. boydii, 172
S. dysenteriae, 165, 172
S. flexneri, 172
S. sonnei, 172
Silage, 87
Silent mutations, 110
Similarity matrix, 31
Single cell isolation, 186
Single cell protein (SCP), 149, 151, 158
Skin flora, 177
Sleeping sickness, 175
Slime moulds, 23
Smallpox, 2, 166
Snapping division, 36
Soil microorganisms, 136
Specialized transduction, 119
Spikes, 43, 49
Spirilla, 30, 35
Spirillaceae, 35
Spirochaetes, 32
Spiroplasmas, 18
Spirulina, 33, 37, 158
Spoilage of foods, 162
Sporocytophaga, 33, 38
Sporolactobacillus, 86
Sporothrix, 174
Sporozoa, 23
Staining procedures, 187
Staphylococcus, 32, 35, 163, 166, 167, 168
S. aureus, 36, 115, 129, 130, 169, 177
Starch, 145
Starter cultures, 144
Stemphylium, 160
Sterilization-dry, 185
Sterilization, (wet) moist, 185
Steroids, 158
Stigonema, 33, 37
Stop codon, 111
Strain development, 151
Streak plate, 186, 187
Streptococcus, 32, 36, 86, 160, 163, 166, 167, 168
S. cremoris, 86, 158, 159
S. lactis, 158, 159

S. mutans, 166
S. pneumoniae, 117, 165, 168
S. pyogenes, 36, 169, 170, 177
S. sobrinus, 166
S. thermophilus, 36, 159
Streptomyces, 33, 38, 39, 154, 157
S. albus, 154
S. aureofaciens, 147
S. erythreus, 147, 157
S. fradiae, 157
S. griseus, 147, 157
S. kanamyceticus, 157
S. rimosus, 147, 157
Streptomycin, 157, 171
Streptosporangium, 33, 40
Stroma, 13
Subcellualr organelles, 12
Subcutaneous mycoses, 174
Succinate, 78, 82, 99
Sulphonamides, 180
Sulphate reduction, 91
Sulfolobus, 33, 41, 102, 103
Sulphur cycle, 140
Sulphur oxidizing
  bacteria, 38
  organisms, 101
Sulphur-reducing bacteria, 35, 83
Sulphur springs, 38, 102, 142
Superficial mycoses, 174
Superoxide, 131
Synchytrium endobioticum, 26
Synchronous growth, 67
Syphilis, 174, 178

T

T-cells, 126, 127, 129, 131
TCA cylce (Tricarboxylic acid cycle), 74, 79, 80, 81
t-DNA, 123
Target site, 117
Tc-cells, 131
Teichoic acids, 8, 35
Temperate phage, 58
Terminal electron acceptor, 70, 139
Testa, 21
Tetanus, 173
Tetanus antitoxin, 173
Tetanospasmin, 173
Tetracyclines, 157
TH cells, 127, 131
Thermoactinomyces, 33, 40
Thermocline, 142
Thermomonospora, 33, 39, 40
Thermophiles, 41, 65
Thermoplasma, 33, 41
Thermoproteus, 33, 41
Thermothrix, 102
Thermotolerant, 65
Thiobacillus, 33, 38
T. denitrificans, 38, 102
T. ferrooxidans, 38, 76, 102, 103, 141
T. intermedius, 76
T. novellus, 38, 81
T. thiooxidans, 38, 65, 102, 141
T. thioparus. 102, 141
Thiomicrospora, 33, 38
Thiospirillum, 33
Thiothrix, 141
Thylakoid, 21, 37
Thymus, 125
Tinsel type, 15
TI plasmid, 34, 115
TMV virion, 49
Torula, 160
Torulopsis, 160
Toxins, 168
Toxoid, 170, 173
Transcription, 95, 111, 113
Transducing phage, 119
Transduction, generalized, 119
Transfer RNA (tRNA), 96, 111
Transformation, 117
Translation, 95, 112, 113, 114
Transposable pahge, 59
Transposons, 115, 116
Treponema, 32, 167
T. pallidum, 34, 165, 174
2, 4, 5-Trichlorophenoxy acetic acid, 143
Trichomes, 30
Trichophyton, 174, 175
Trichothecens, 164
Trickling filters, 143
Trimethoprim, 180
Trypan blue, 182
Trypanosoma gambiense, 21
T. brucci, 131
Triplet code, 96, 110
Tryptophan, 120
Tuberculous meningitis, 171
Tuberculin test, 170
Tuberculosis, 170, 171, 178
Turbidity method, 66

Turbidostat, 149
Typhoid fever, 171

U

Ubiquinone, 71
UDP galactose, 93
Unit membrane, 8
Ustilago, 26

V

Vaccines, 132
Vectors, 121, 178
Venturia inaequalis, 26
Vibrio, 160, 163, 167, 171
V. anguillarum, 172
V. cholerae, 35, 130, 165, 167, 168, 169, 173
V. parahaemolyticus, 172
Vibrionaceae, 35
Vibrios, 30
Vinegar, 88
Virion, 42
Viroids, 17
Virulence plasmids, 114
Virulent phage, 58
Viruses (and virus groups), 18, 42, 49
- and cancer, 60
- caused by, 42
- entry into host, 48
- envelope, 42
- genome, maps of, 57
- laboratory study of, 49
- purification of, 50
- symmetry of, 42
- transmission of, 44
  Adenoviridae, 46, 54
  Arbovirus group A, 52
  ARD virus, 46, 54
  Caulimoviridae, 47, 52
  Cauliflower mosaic virus, 47
  Citrus exocortis, 17
  Comoviridae, 51
  Corona virus, 53
  Cowpea mosaic virus, 51
  European encephalitis virus, 52
  Epstein-barr virus, 60
  Gemini virus, 47
  Hepatitis B virus, 60
  Herpes simplex virus, 46, 54
  Herpesviridae, 46, 54, 60
  Human immunodeficient virus (HIV), 46, 55, 131, 134
  M13, 45, 56
  Influenza virus, 46, 53
  HTLV-1, 60
  HTLV-2, 60
  Maize streak virus, 47
  Measles virus, 53
  MS 2, 45, 56, 57
  Mumps virus, 53
  Mu, 59
  Nepoviridae, 47
  Orthomyxoviridae, 46, 53
  Papovaviridae, 46, 54
  Paramyxoviridae, 53
  Parvoviridae, 46, 54
  Parvovirus, 46
  Pf, 58
  Phage lambda (λ), 45, 58, 59
  ϕ6, 45, 60
  ϕ×174, 45, 56, 57
  Polio virus, 45, 52
  Polyoma virus, 54
  Potato yellow dwarf, 47
  Potato virus X (PVX), 19, 47, 51
  Poxviridae, 46, 54
  Protexviridae, 47
  Rabies virus, 45, 53
  Reoviridae, 46, 54
  Reovirus, 46
  Rhabdoviridae, 47, 53
  Retroviridae, 46, 55
  Rice dwarf virus, 47
  Satellite virus, 52
  Simean vacuolating virus, 60
  SV40, 46, 54, 60
  T2, 118
  T4, 45, 58
  T7, 57, 58
  Tobacco mosaic virus (TMV), 47, 51
  Tobacco necrosis virus, 52
  Tobamoviridae, 47
  Tobacco rattle virus, 47, 51
  Tobacco ringspot virus, 47
  Tobraviridae, 47
  Tomato bushy stunt virus, 51
  Tomato spotted wilt virus, 47
  Tombusviridae, 51
  Vaccinia virus, 46, 54
Vitamins, 64, 155

Volutin granules, 11
Volvox, 24

W

Whey, 159
Wild strains, 108
Wine making, 84
Whiplash type, 15
Whooping cough, 171

X

Xanthomonas phaseoli, 76
Xenobiotic, 142
Xylulose-6-phosphate, 80

Y

Yeasts, 84, 145, 158
Yersinia, 172
Yersinia pestis, 32, 34, 165, 167, 172
Yogurt, 159

Z

Zearalinone, 164
Zooplankton, 142
Zygomycetes, 26
Zygote, 16
Zymomonas, 84